T0332096

A Gallery of Combustion and Fire

A Gallery of Combustion and Fire is the first book to provide a graphical perspective of the extremely visual phenomenon of combustion in full color. It is designed primarily to be used in parallel with, and supplement existing combustion textbooks that are usually in black and white, making it a challenge to visualize such a graphic phenomenon. Each image includes a description of how it was generated, which is detailed enough for the expert but simple enough for the novice. Processes range from small-scale academic flames up to full-scale industrial flames under a wide range of conditions such as low and normal gravity, atmospheric to high pressures, actual and simulated flames, and controlled and uncontrolled flames. Containing over 500 color images, with over 230 contributors from over 75 organizations, this volume is a valuable asset for experts and novices alike.

DR. CHARLES E. BAUKAL, JR. is Director of the John Zink Institute, a training and educational organization for combustion engineers and researchers using John Zink Co. LLC equipment, former Chair of the Central States Section of the Combustion Institute, and Program Chair for the American Society for Engineering Education.

DR. AJAY K. AGRAWAL is the Robert F. Barfield Endowed Chair in Mechanical Engineering at the University of Alabama.

DR. SANDRA OLSON is from the NASA Glenn Research Center and has served as Primary Investigator, Co-Investigator, and Project Scientist on numerous flight experiments and ground-based studies.

DR. MICHAEL J. GOLLNER is an assistant professor and Deb Faculty Fellow in the Department of Mechanical Engineering at the University of California, Berkeley.

DR. TIMOTHY J. JACOBS is a professor and the Steve Brauer Jr. '02 Faculty Fellow in the Department of Mechanical Engineering at Texas A&M University.

DR. MARK VACCARI is a development engineer at John Zink Hamworthy Combustion and an adjunct professor at the University of Tulsa.

A Gallery of Combustion and Fire

Edited by

Charles E. Baukal, Jr.

John Zink Co. LLC

Ajay K. Agrawal

University of Alabama

Sandra Olson

NASA Glenn

Michael J. Gollner

University of California, Berkeley

Timothy J. Jacobs

Texas A&M University

Mark Vaccari

John Zink Hamworthy Combustion

CAMBRIDGE
UNIVERSITY PRESS

CAMBRIDGE
UNIVERSITY PRESS

University Printing House, Cambridge CB2 8BS, United Kingdom

One Liberty Plaza, 20th Floor, New York, NY 10006, USA

477 Williamstown Road, Port Melbourne, VIC 3207, Australia

314–321, 3rd Floor, Plot 3, Splendor Forum, Jasola District Centre,
New Delhi – 110025, India

79 Anson Road, #06–04/06, Singapore 079906

Cambridge University Press is part of the University of Cambridge.

It furthers the University's mission by disseminating knowledge in the pursuit of education, learning, and research at the highest international levels of excellence.

www.cambridge.org
Information on this title: www.cambridge.org/9781107154971
DOI: 10.1017/9781316651209

First published 2020

Printed in Singapore by Markono Print Media, Private Limited

A catalogue record for this publication is available from the British Library.

Library of Congress Cataloging-in-Publication Data
Names: Baukal, Charles E., Jr., 1959– editor.
Title: A gallery of combustion and fire / Charles E. Baukal, Jr. (John Zink Co., LLC) [and five others].
Description: Cambridge ; New York, NY : Cambridge University Press, 2019. | Includes bibliographical references and index.
Identifiers: LCCN 2019019428 | ISBN 9781107154971 (hardback : alk. paper)
Subjects: LCSH: Flame–Pictorial works. | Combustion–Pictorial works. | Fire–Pictorial works.
Classification: LCC QD516 .G245 2019 | DDC 341/.361–dc23
LC record available at https://lccn.loc.gov/2019019428

ISBN 978-1-107-15497-1 Hardback

Contents

Preface

This book is the result of two primary factors. The first is the success of a previous Cambridge University Press book published in 2003 entitled *A Gallery of Fluid Motion* edited by M. Samimy, K. S. Breuer, L. G. Leal, and P. H. Steen. That book was patterned after two earlier graphical books on the subject of fluid flow. The first is *Illustrated Experiments in Fluid Mechanics: The NCFMF Book of Film Notes* by Asher H. Shapiro, published by MIT Press in 1972. The second was a very successful book by Milton Van Dyke, entitled *An Album of Fluid Motion*, published by Parabolic Press in 1982. All of the images that appear in *A Gallery of Fluid Motion* were winning images as selected by the American Physical Society in an annual contest held since 1983. The mostly color images were selected for artistic and technical merit. All of the images were previously published in the journal *Physics of Fluids*.

The second factor leading to this book is the Combustion Art contest sponsored by the Central States Section of the Combustion Institute (CSSCI). Sandra Olson, one of the editors of the present book, has led this competition for many years where the winning art since 2004 is posted online (www.cssci.org/). Cambridge University Press heard about this competition and contacted the CSSCI about authoring a book similar to *A Gallery of Fluid Motion* but on the subject of combustion. Four of the editors of the current book (Agrawal, Baukal, Jacobs, and Olson) are or were members of the CSSCI board. The initial book title was extended from *A Gallery of Combustion* to *A Gallery of Combustion and Fire* when an extensive section on fire was added. The International Association for Fire Safety Science sponsors a fire science image contest that was of interest to this project.

This is the first known book of its type, as no other books were found that take a primarily graphical approach to combustion. There are a number of unique aspects of the book. There are numerous full-color images of all aspects of combustion including actual and simulated flames and fires. Each image includes a brief description with details of how it was generated. The book includes small-scale academic flames, large industrial flames, and fires. A wide range of pressures is considered, including low gravity, atmospheric, and pressurized flames relevant in space, industrial, and propulsion applications, respectively. The book combines the science and art of combustion.

There are many potential benefits of this book. The full-color images are essential to understand combustion, and represent a significant improvement compared to most earlier books on the subject, which are in black and white. There is a full range of flame types from academic to industrial, whereas most combustion books deal with one or the other. The book includes both traditional controlled flames as well as uncontrolled fires. It is detailed enough for the expert yet simple enough for the novice.

The book is designed primarily to supplement existing combustion textbooks used in both undergraduate and graduate courses as well as in short courses on combustion. It should be of interest to anyone working in the field of combustion, including professors, researchers (academic, industrial, and government), designers, students, industrial users, and even to nontechnical readers interested in this subject. While most of the images are included for pedagogical purposes, there are some of artistic merit as well. Due to space limitations, no attempt has been made to be comprehensive, but an objective of the book is to be representative of the numerous types and aspects of combustion. No prerequisite knowledge is required to use the book.

Introduction

Charles E. Baukal, Jr.

Combustion is a critically important and extremely visual phenomenon. When properly controlled, combustion is important for a wide range of processes. For example, it is the primary source of power generation for our vast array of electrical equipment and electronics. However, when combustion is not properly controlled, it can be a source of great devastation. For example, uncontrolled wildfires are still a major concern in many parts of the world.

While knowledge continues to expand on the subject of combustion, there is still much that remains to be learned. Considerable research continues on reducing pollution emissions and increasing thermal efficiencies from combustion processes. However, combustion continues to be an extremely complex phenomenon. The fluid flow is typically turbulent, which is difficult to model. The heat transfer is nonlinear due to the importance of thermal radiation. To further complicate matters, the radiation is spectrally dependent due to the effects of water vapor and carbon dioxide in the combustion products. Combustion chemistry is extremely complicated, where the reaction of a relatively simple fuel like methane with air consists of 325 reactions and 53 species, most of which last only a fraction of a second in the flame [1]. In many cases, the rate constants for the numerous reactions are not well known, which makes modeling that much more difficult. There may be acoustic and pressure effects in certain types of combustion processes. There may also be a wide disparity in the length scales in some systems where, for example, the fuel injectors may have very small holes while the flames may be inside very large furnaces. When these systems are discretized for numerical modeling, there may be tens of millions of control volumes. In some systems, the combustion process is time dependent, which greatly increases the computational requirements. This means very large computational power and storage may be required for modeling many important types of combustion processes.

Since computational models are not currently adequate to accurately predict all aspects of combustion, experiments continue to be important in furthering the understanding of the physics of the many processes in combustion. One of the important outputs from these experiments is often graphic images that can greatly add to understanding those processes and can also be invaluable in validating computational models. Both aspects of the images are of interest here.

Flames may exhibit a range of colors such as yellow, red, and/or blue depending on factors such as the specific combustion process, the fuel, and the operating conditions. For example, it is possible that a given burner can produce yellow or blue flames with the same fuel under fuel-rich or fuel-lean conditions, respectively. This book is a compendium of combustion images accompanied by explanatory text. Low-gravity, atmospheric pressure, and pressurized flames using solid, liquid, and gaseous fuels, ranging from small-scale laboratory up to full-scale industrial flames under controlled and uncontrolled conditions, are included. Photographs, drawings, and computer-generated images, some of which were winners in combustion and fire science art contests, are included.

Most combustion books are published in black and white, which makes it nearly impossible to show the colorful and dynamic nature of flames. It also makes it more difficult to determine what is happening in the combustion process. The purpose of this book is to present color images of a wide range of aspects of combustion. While there are analogous graphical books for fluid flow, no such book was found for combustion. This book is designed to fill that gap. It can also serve as a supplement to existing books describing the combustion phenomenon.

The book is divided into six chapters. Chapter 1 is on combustion fundamentals and is divided into eight sections: simple flames, laboratory and idealized flames, practical flames, spherical flames, gas jet and liquid fuel flames, flames at high speed, coal and solid-particle combustion, and combustion of metals. These are generally smaller-scale combustion processes studied in great detail in the laboratory.

Chapter 2 contains a range of images concerning the numerical simulation of flames commonly referred to as computational fluid dynamic (CFD) modeling. Some humorously refer to CFD as *colorized* fluid dynamics because of the impressive graphics that can be generated from a simulation. Because of advancements in numerical techniques, combustion physics, and computer hardware, CFD has become a standard tool for designing and analyzing combustion processes. In CFD, the challenge when generating images can be what to include and what to exclude because of the numerous combinations of variables that can be plotted. A wide range of flow types and analysis techniques are displayed in this chapter.

Chapter 3 includes images related to internal combustion engines and turbines used in cars, trucks, and airplanes. These are inherently more difficult combustion processes to visualize because the flames are normally contained inside pressurized metal combustion chambers and are very transient in nature. Significant advancements have been made in optical access to generate images that can provide both qualitative information and quantitative measurements in these systems.

Chapter 4 concerns low-pressure combustion processes that are important in space applications. While removing gravity may simplify the combustion in some aspects, making experimental measurements and producing images are more complicated due to the challenge of either creating low-gravity environments on earth or conducting experiments in space. This is an active area of combustion research, and a wide range of flames is shown in this chapter, which has been divided into the following sections: gaseous fuels, liquid fuels, and both thick and thin solid fuels.

Industrial flames are the subject of Chapter 5. These are divided into six sections: metals industry, process heating, power generation, infrared heating and drying, flares, and oxygen-enhanced flames. Some of these are very large-scale flames; for example, flare flames could be 100 m or more in length. Conducting experiments on these types of systems can be challenging because of the scale involved and because access can be difficult, particularly if there is a possibility of adversely affecting whatever is being produced with the combustion system. The section on oxygen-enhanced flames shows examples of how a traditional combustion process using air as the oxidizer may be improved by using an oxidizer with a higher concentration of oxygen up to using pure oxygen for combustion.

The final chapter contains a wide range of images related to fire. Some distinguish between a flame that is considered to be a controlled combustion reaction and a fire that is often considered to be an uncontrolled combustion reaction. Fires can range from very small-scale like those in Chapter 1 up to very large-scale like those in Chapter 5. A wide variety of materials with varying degrees of flammability may be unintended "fuels" in fires. Fire suppression is also considered. This chapter has sections on pool fires, flame spread and fire growth, fire suppression, fire whirls, wildland fires, and smoldering combustion.

Reference

1. Stephen R. Turns, *An Introduction to Combustion: Concepts and Applications*, 3rd ed., McGraw-Hill, New York, 2011.

1 Fundamental Flames

Edited by Ajay Agrawal

Introduction

Ajay Agrawal

Combustion is ubiquitous around us and in the technology that we depend on in our daily lives. To someone uninitiated in the field, observing the combustion process can produce emotions ranging from apathy, fear, boredom, conservatism, or caution to outright joy, including love and affection. Observing combustion is also a great source of fascination to many, in particular to scientists and engineers, who seek to understand the intricate relationships among the chemical, thermal, fluid dynamic, and other complex phenomena involved in the combustion process. The wide variety of fuel and oxidizers, operating conditions such as initial temperature, pressure, and concentrations, and combustor geometry configurations create virtually limitless combinations of combustion, observed visually as flame(s). For example, flames of gaseous, liquid, and solid fuels share many similarities, but they also exhibit their own unique features. Flames do not depict all of their features naturally, and thus researchers employ various diagnostics techniques, often laser-based, to measure one or more characteristics, which can also unravel the mysteries of soot, nitric oxide, and carbon oxide formation, and/or help address various safety-related issues such as flame flashback, flame blow-off, flame noise, etc. The ultimate goal of such explorations is to develop clean combustion systems that are fuel agnostic, highly efficient, and, most importantly, environmentally friendly, i.e., they produce low or very low levels of harmful emissions, including greenhouse gas emissions. This chapter presents simple canonical flames, free from the complexities of most practical combustion systems, to offer insight into the key phenomena. This chapter is divided into eight sections, each covering a different aspect of laboratory-scale flames.

- The first section contains images of (1) laminar diffusion flames to illustrate how shape and sooting tendencies are affected by the oxidizer flow rate in a simple Burke–Schuman burner as shown in Figure 1.1; (2) a premixed methane–air flame exhibiting blue color distinct from the yellow color of diffusion flames as shown in Figure 1.2; and (3) a triple flame comprised of both diffusion and premixed flames in a single flame. The flames widen when gravity-induced buoyancy is removed, as shown in Figures 1.3 and 1.4. Figure 1.5 shows a well-defined conical reaction zone in a premixed flame, which can become corrugated by the presence of turbulence in the reactant flow. Images illustrate various phenomena captured by multiple observational techniques, including shadowgraph and color schlieren.
- The images in Section 1.2 show one of the simplest flame configuration, the so-called one-dimensional flame obtained in a counterflow burner configuration. Images show how flame thickness, flame position, flame color, and/or flame luminosity are all affected by the oxidizer, equivalence ratio, hydrogen enrichment, strain rates, etc. Tulip flames obtained experimentally and computationally are shown in Figure 1.12. Transition in turbulent flames and flame propagation in narrow channels are illustrated next.
- Figures 1.15 and 1.16 show vortical structures associated with noise generation in flames. The flame structure and its noise characteristics can be improved by introducing a porous structure within the combustor as shown in Figure 1.17.
- Section 1.4 presents spherical flames created by locally igniting the reactant mixture. As the flame propagates radially, and the flame front increases in size, the cellular instabilities are formed and can be clearly observed.
- Section 1.5 presents examples of liquid flames, produced either by injecting the fuel into the oxidizer or by allowing the fuel to find the oxidizer, such as in pool fires.
- Section 1.6 presents an example of combustion at supersonic conditions.
- Combustion of coal particles, one of the most prevalent solid fuel in use, is illustrated in Section 1.7. Coal can be burned alone or in combination with other gaseous fuels. Figures 1.26–1.28 show that metals can also burn when introduced in typical methane–air flames, and produces interesting features.
- Section 1.8 shows that different metal powders and/or metals burning in air exhibit similarity to gaseous fuel combustion.

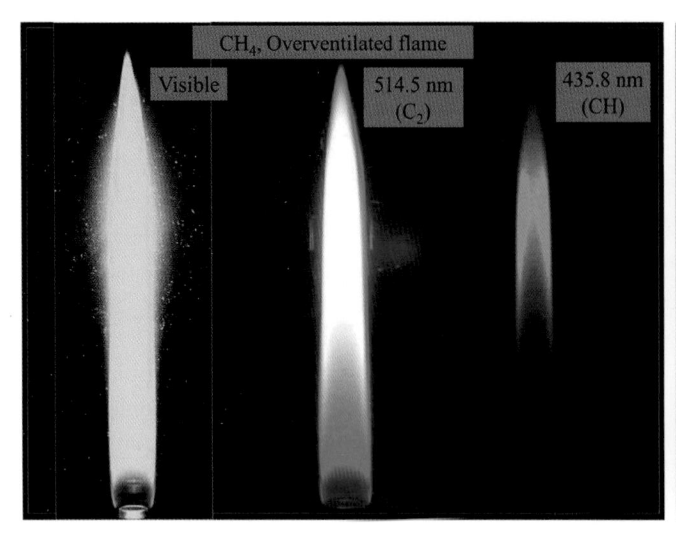

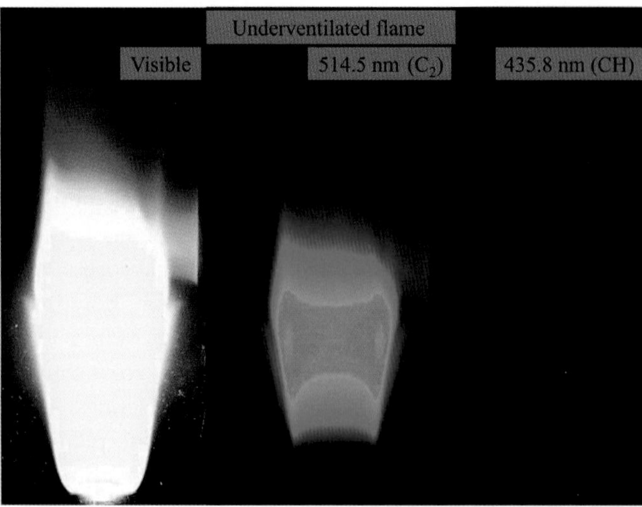

Figure 1.1

1.1 Different Types of Simple Flames

1.1.1 Overventilated and Underventilated Axisymmetric Laminar Diffusion Flames in a Burke–Schumann Burner

Loreto Pizzuti and Fernando de Souza Costa
INPE National Institute for Space Research

Overventilated (left) and underventilated (right) laminar diffusion flames are shown. Overventilated flames are closed on the axis of symmetry of the burner while underventilated flames are open and terminate over the external oxidant tube. The two types of flame can be obtained keeping the flow momentum of the reactants equal and by changing the fuel tube radius, a, the Peclet number, Pe, or the initial mass fraction of oxygen in air, YO0. Furthermore, they can be obtained by varying the mass flow rates of one of the reactants and keeping constant the flow rate of the other.

The left image shows an overventilated CH_4/air flame, with a = 3 mm and fuel velocity of 550 mm/s obtained in the visible spectrum as well as with optical filters for C_2 and CH radicals detection. The right image shows an underventilated CH_4/air flame obtained by keeping constant the mass flow rate of CH_4 while reducing the mass flow rate of air in the visible spectrum and with optical filters for C_2 and CH radicals detection.

This research was partially funded by CAPES Brazilian Research Support Agency.

Keywords:
Overventilated; underventilated; axisymmetric; laminar; diffusion; Burke–Schumann burner

Reference

L. Pizzuti and F.S. Costa, "Influence of refraction and divergence of light on tomography of axisymmetric laminar diffusion flames," Fuel, Vol. 106, pp. 372–379, 2013.

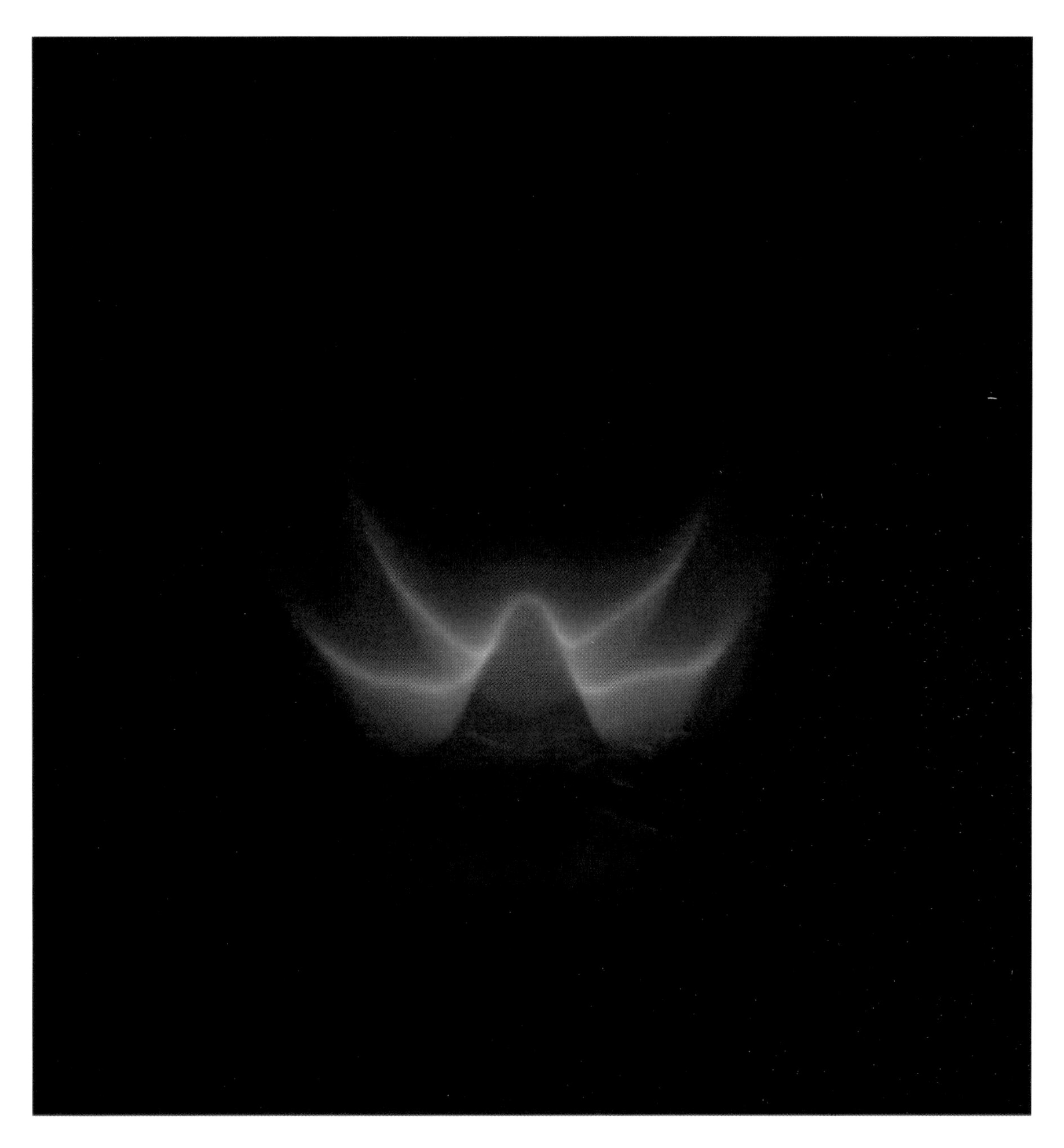

Figure 1.2

1.1.2 Crossed Slot Burner – Premixed Methane Air – Near Blow-off

Scott R. Rockwell

University of North Carolina, Charlotte

This image is of a premixed methane air flame on a steady-state slot burner. The slots are in the shape of a cross on top of a circular cap. The equivalence ratio is near stoichiometric, and the flow rate is high enough that the edges of the flame lift off of the burner while the center core remains anchored.

Keywords:

Premixed; methane/air; crossed slot burner; lift-off

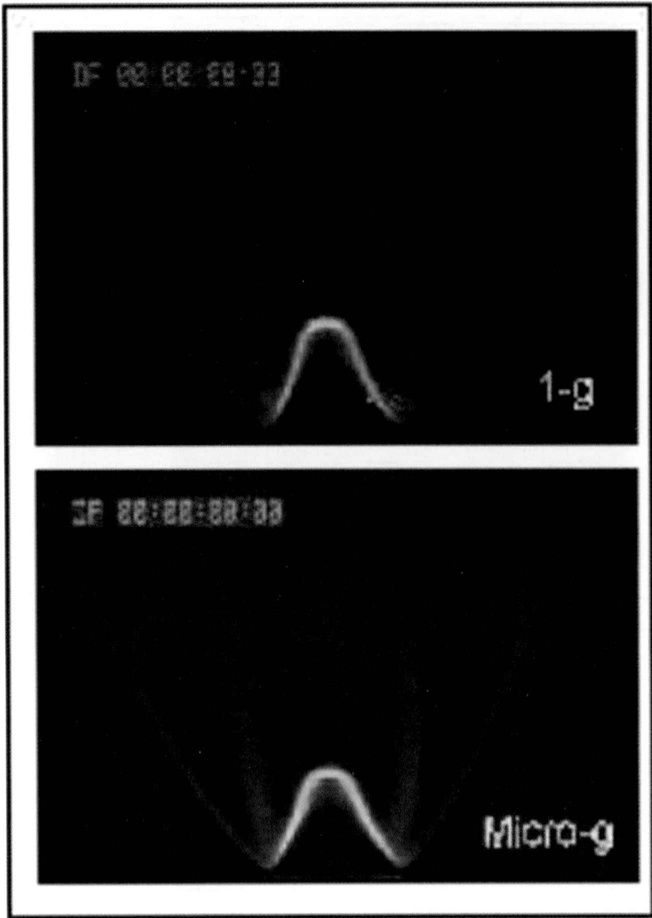

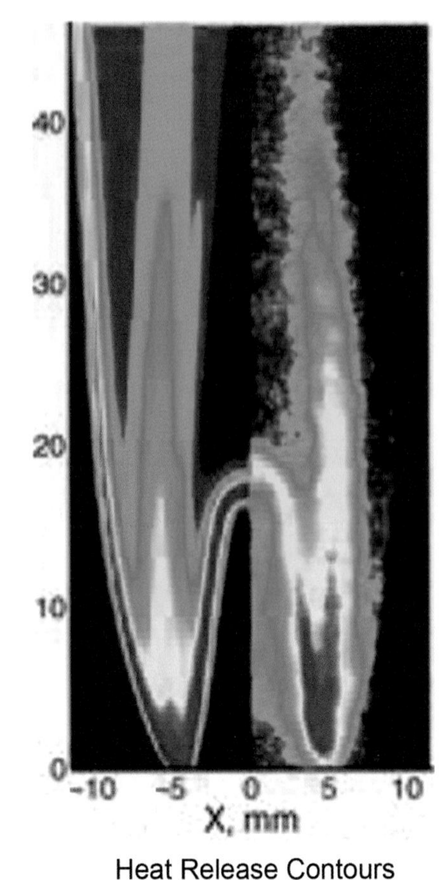

Figure 1.3

1.1.3 Laminar Triple Flames at Normal Gravity and Microgravity

Suresh K. Aggarwal and Ishwar K. Puri
University of Illinois, Chicago

The left-side contains direct images of methane triple flames under normal (1g) and microgravity (μg) established on a Wolfhard–Parker slot burner by introducing a fuel rich methane-air mixture from an inner slot and a fuel lean mixture from two symmetric outer slots. A triple flame is characterized by three reaction zones. Two premixed reaction zones (one fuel–rich on the inside and the other fuel–lean on the outside) form the exterior "wings"of the flame, and a nonpremixed reaction zone is established in between the two wings. While the inner rich premixed zone is only weakly affected by gravity, the outer lean premixed zone and the nonpremixed zone are noticeably influenced by buoyancy.

The right-side image compares the computed and measured triple flames in terms of the computed volumetric heat release rates and the experimentally measured C2*-chemiluminescence intensities. While all three reaction zones are clearly visible in the computed heat release rate contours, the lean premixed reaction zone is not quite as evident in the experimental image, since the fuel lean chemistry is not captured by C2*-chemiluminescence.

This research was partially funded by National Science Foundation (NSF) and by NASA Microgravity Combustion program.

Keywords:
Laminar Triple Flame; microgravity

Reference

S.K. Aggarwal, "Extinction of Laminar Partially Premixed Flames," Progress in Energy & Combustion Science, Vol. 35, pp. 528–570, 2009.

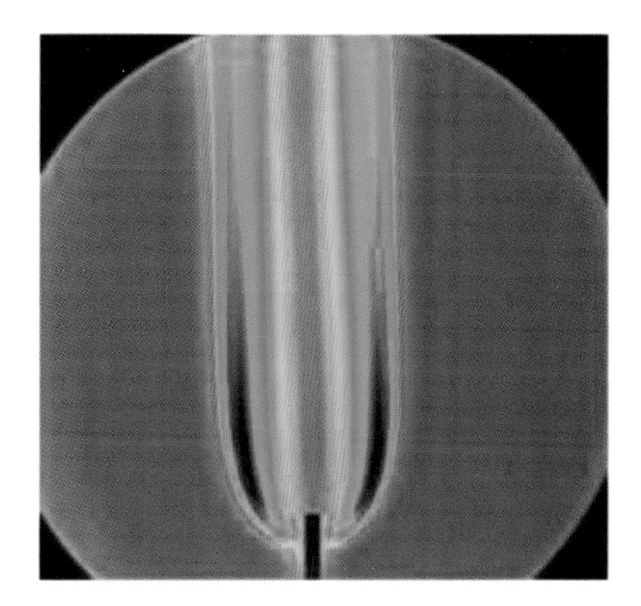

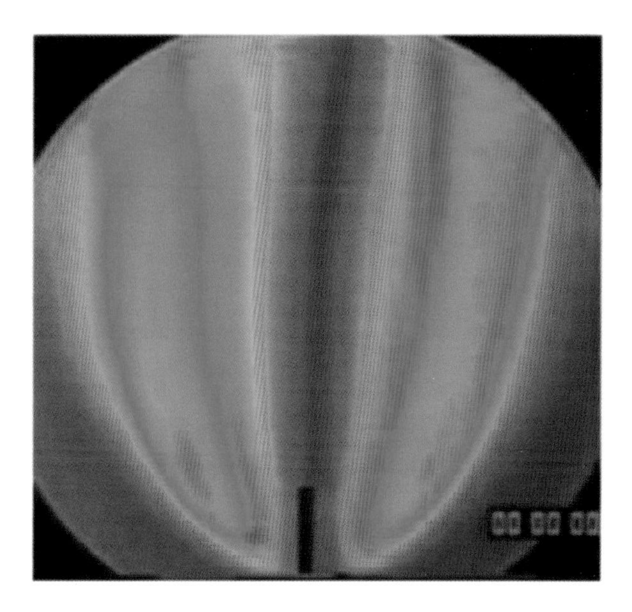

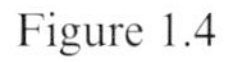
Figure 1.4

1.1.4 Rainbow Schlieren Image of Laminar Hydrogen Jet Diffusion Flame in Normal Gravity (Left) and Microgravity (Right)

Khalid Al-Ammar and Ajay K. Agrawal

University of Alabama

Rainbow schlieren images of laminar diffusion flames of hydrogen issuing from a small circular tube of (1.2 mm diameter) are shown in normal gravity (left) and microgravity (right). The fuel jet Reynolds number is about 100. Images were obtained in the 2.2 s drop tower at NASA Glenn Research Center. Hydrogen flames are visually transparent, but they can be detected using the schlieren technique which is sensitive to density gradients in the flow field. In rainbow schlieren, the density gradients are color-coded by replacing the knife-edge by a continuously graded color filter.

Gravity has a significant effect on the flame structure. In normal gravity, the flame is narrow with large color variations indicating large density gradients. In microgravity, the flame is much wider with gradual color changes indicating smaller density gradients. The images can be analyzed to infer the local temperature field across the whole field of view.

This research was partially funded by NASA Microgravity Combustion program.

Keywords:
Rainbow Schlieren; laminar; hydrogen; diffusion; microgravity

Reference

Al-Ammar, K.N., Agrawal, A.K. and Gollahalli, S.R., 2000. Quantitative measurements of laminar hydrogen gas-jet diffusion flames in a 2.2 s drop tower. Proceedings of the Combustion Institute, 28(2), pp. 1997–2004.

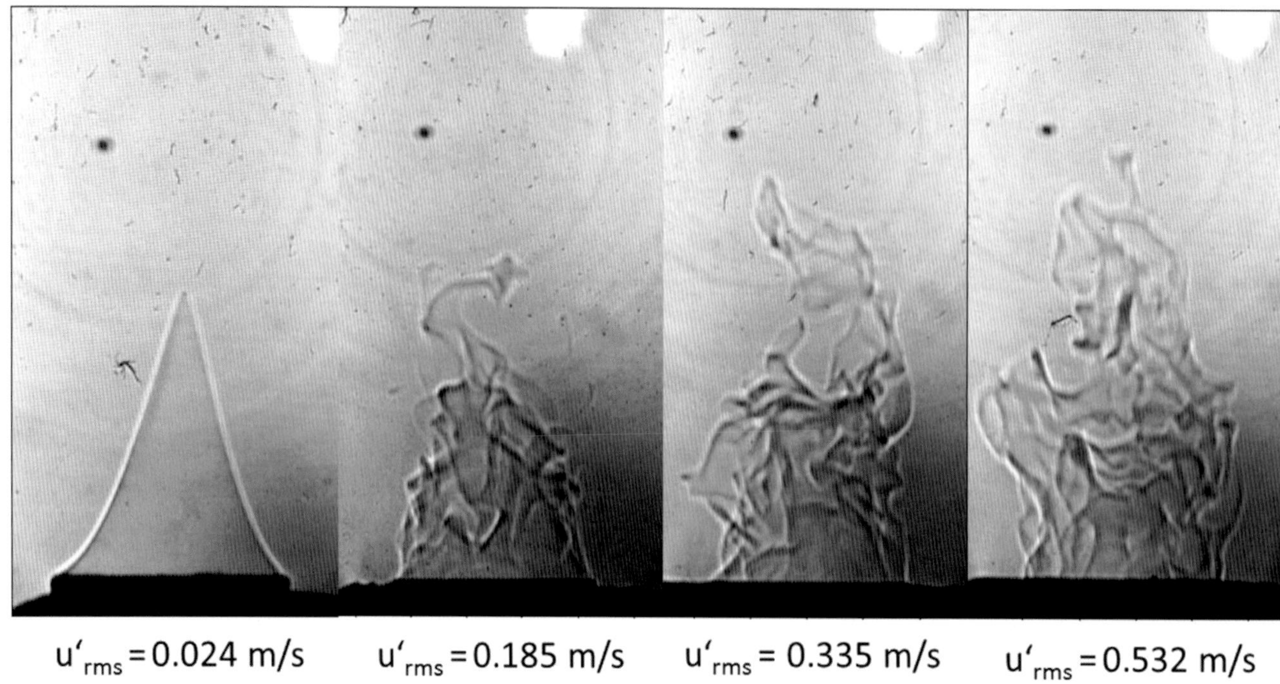

Figure 1.5

1.1.5 Shadowgraph of Methane Air Flames at Various Turbulent Intensities

Scott R. Rockwell

University of North Carolina, Charlotte

Keywords:

Methane; shadowgraph; turbulent

Shadowgraph images of premixed methane air flames on a stabilized burner with equivalence ratio of 1.0 and increasing turbulent intensities ranging from 0.024 m/s to 0.532 m/s are shown.

This research was partially funded by NSF GRFP award.

Reference

Rockwell, Scott R., Rangwala, Ali, S., Influence of Coal Dust on Premixed Turbulent Methane-Air Flames, Combustion and Flame 2013, Vol.160, (3) 635–640.

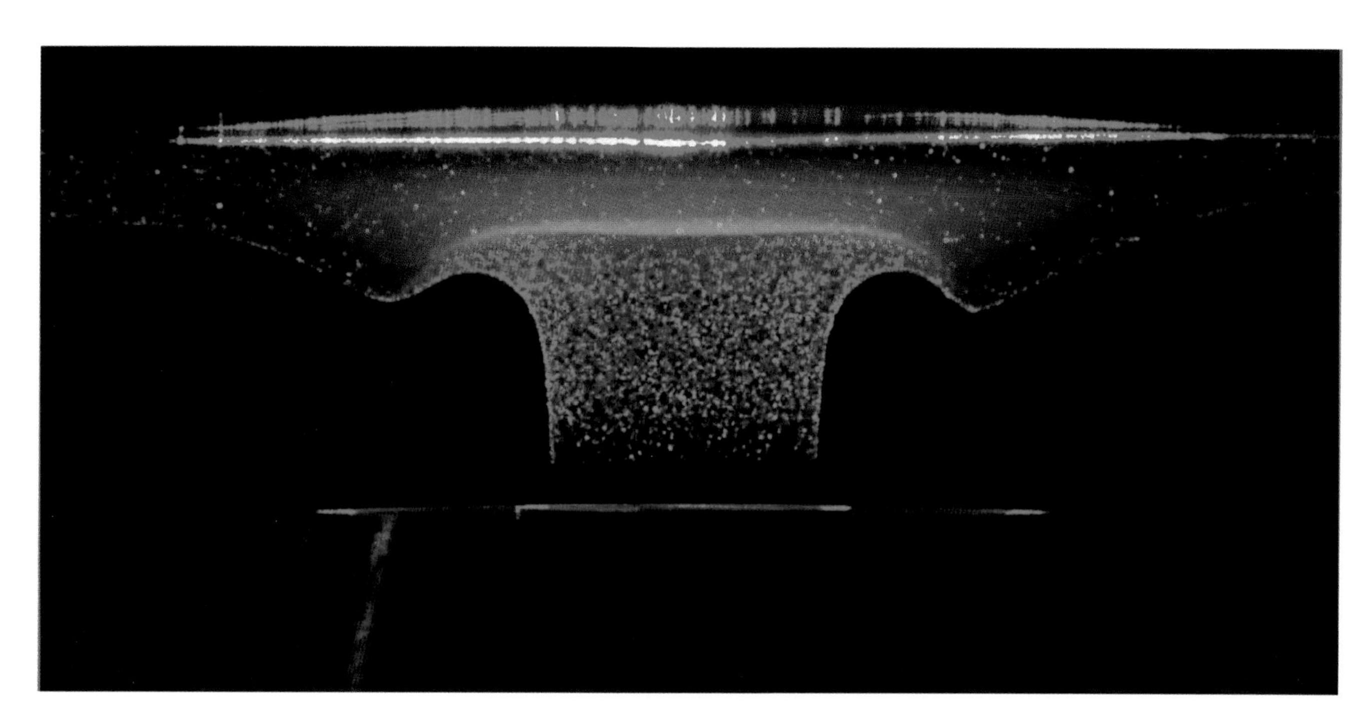

Figure 1.6

1.2 Laboratory and/or Idealized Flames

1.2.1 Laminar Methane Stagnation Flame with Seeding Particles Illuminated by a Laser Sheet

Sean D. Salusbury and Jeffrey Bergthorson
McGill University

Laminar methane-air flame stabilized in a jet-wall stagnation flow, with the flow issuing from the nozzle visible at the bottom of the image. The stagnation flame is visible because of chemiluminescence from the methylidine radical, CH, at wavelengths near 430 nm in the violet part of the visible spectrum. Micron-sized aluminum oxide particles are seeded in the flow and illuminated with a Nd: YAG laser at 532 nm. The laser also illuminates the water-cooled stagnation plate at the top of the image and the fuel-air and co-flow nozzles at the bottom of the image. The seeding density in this image is much higher than would be required for particle-image velocimetry or particle-tracking velocimetry diagnostics, in order to clearly visualize the flow field. Blackbody radiation from the inert seeding particles in the post-flame zone shows the shape of the high-temperature zone and the drop in the gas temperature towards the cooled stagnation plate is also visible.

Keywords:
Methane; stagnation flame; laminar

Reference

https://commons.wikimedia.org/wiki/File:PIV_through_stagnation_flame.jpg

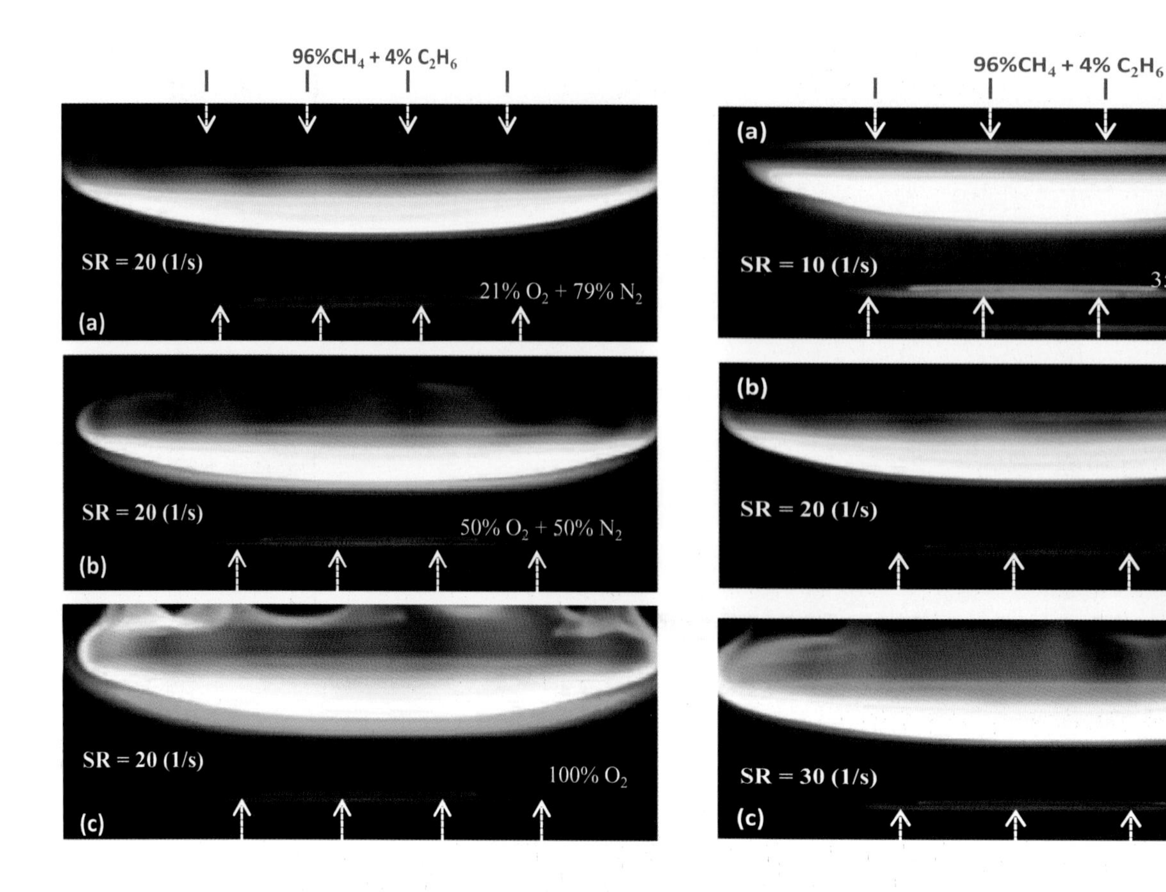

Image I: Effect of oxygen in a counter-flow flame.

Image II: Effect of strain rate in a counter-flow flame

Figure 1.7

1.2.2 Flow Visualization in an Oxy/Fuel Counterflow Burner

Courtney Baukal, Alireza Abdihamzehkolaei, Walmy Cuello Jimenez, and Wilson Merchan-Merchan

University of Oklahoma

Alexei V. Saveliev

North Carolina State University

Flow visualization experiments (FVEs) were implemented in a counter-flow burner to provide visible flow patterns generating a profile in the empty space between the two opposing nozzles that resembles a typical 1-D flame. In this work FVEs are conducted by introducing solid spherical particles in the air flow and by illuminating them as they are carried by the air flow in the space between the burner nozzles.

As the oxygen content in the oxidizer stream was increased, the blue zone became thicker while the overall flame was compressed (Image-F). The maximum temperature and the stagnation plane shifted towards the fuel side and the high temperature region was expanded as the oxygen content was increased. Image-G shows the flame structure as the strain rate is varied for fixed oxygen content of 35%. The strain rate of $10s^{-1}$ has the largest yellow zone; as the strain rate increased, the yellow zone decreased. The blue zone and the overall thickness of the flame also decreased with an increasing strain rate.

This work was supported by the National Science Foundation.

Keywords:

Oxy/fuel; counterflow flame

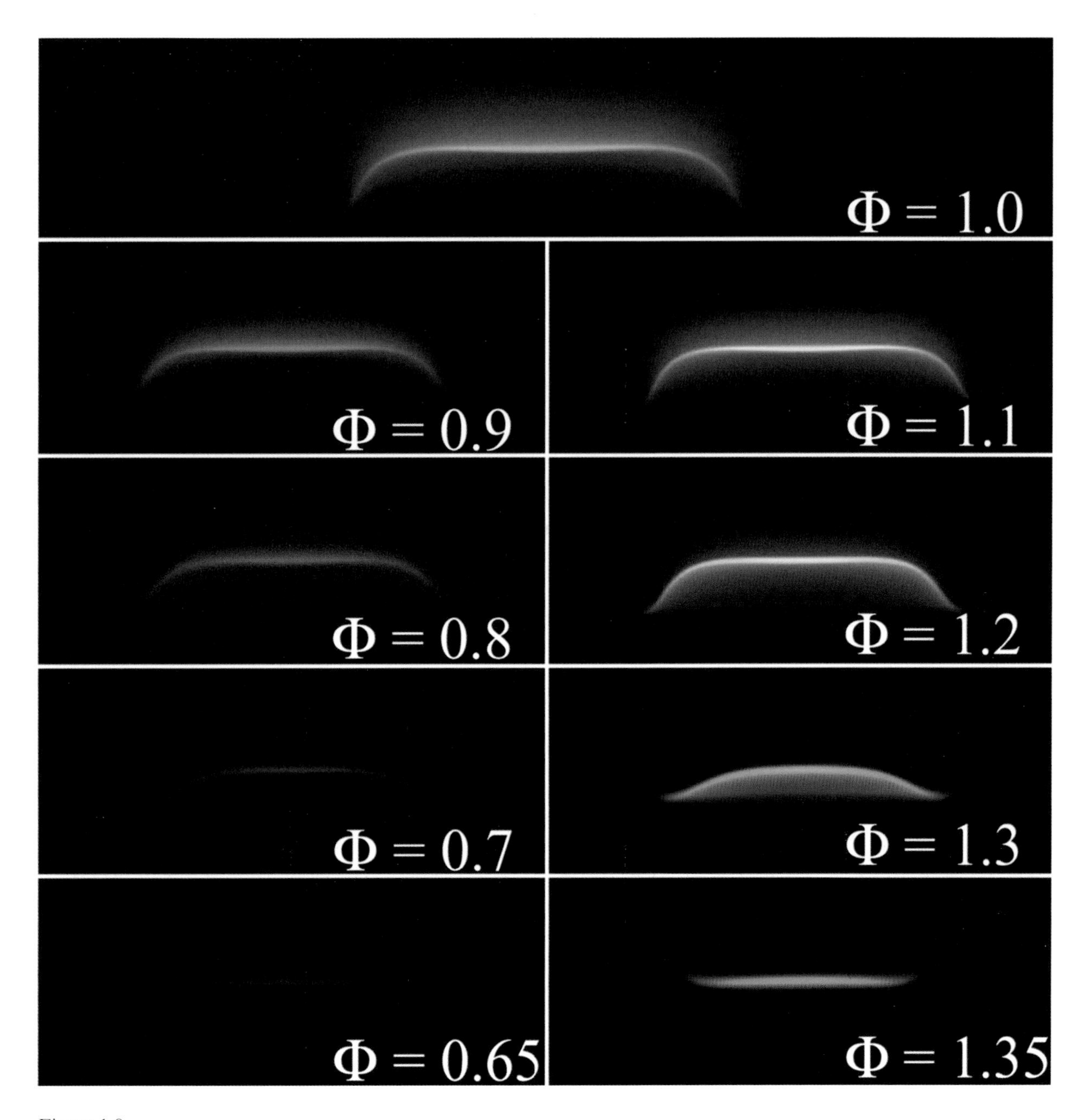

Figure 1.8

1.2.3 Laminar Methane Stagnation Flames at Variable Equivalence Ratios

Jeffrey M. Bergthorson and Paul E. Dimotakis

California Institute of Technology

Laminar methane-air flames stabilized in a jet-wall stagnation flow at varying fuel-air equivalence ratios, Φ. Lean flames, with $\Phi < 1$, have a violet color that is the result of chemiluminescent emission of the methylidine radical, CH, at wavelengths near 430nm. As the flames become fuel-rich, with $\Phi > 1.1$, the flames increasingly turn to a green-blue color that results from emissions of the C_2 Swan bands, which emit light from 430 to 650 nm.

This work was funded by the Air Force Office of Scientific Research under grant F49620-01-1-0006.

Keywords:
Methane; stagnation flames; laminar

Reference

JM Bergthorson. Experiments and Modeling of Impinging Jets and Premixed Hydrocarbon Stagnation Flames. PhD thesis. California Institute of Technology, 2005. http://resolver.caltech.edu/CaltechETD:etd-05242005-165713.

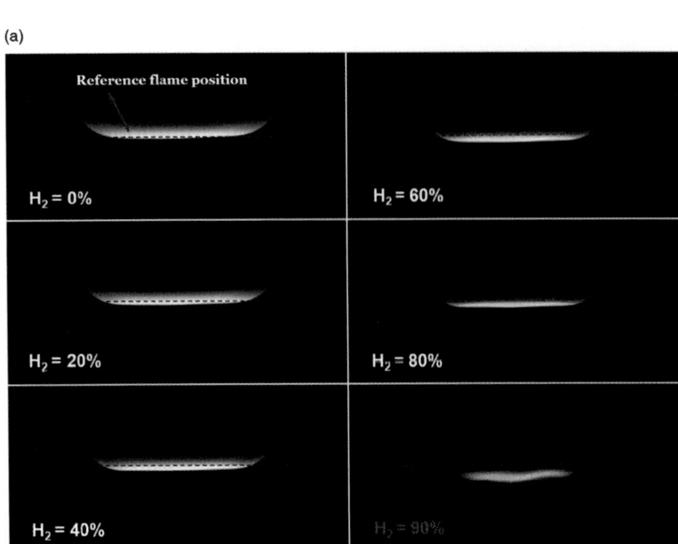

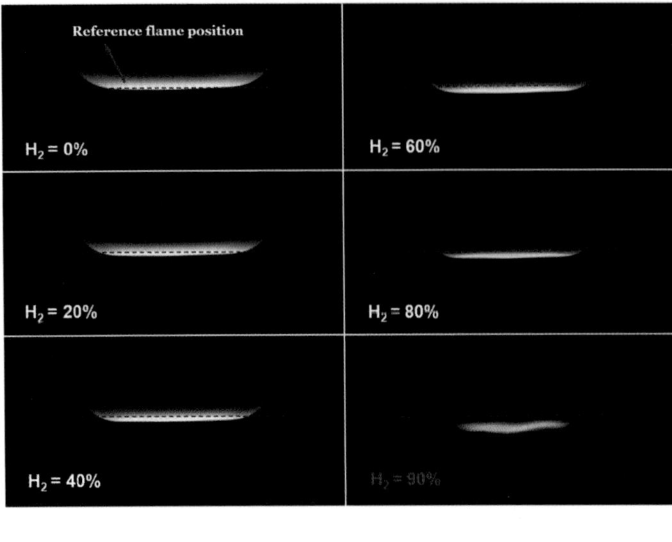

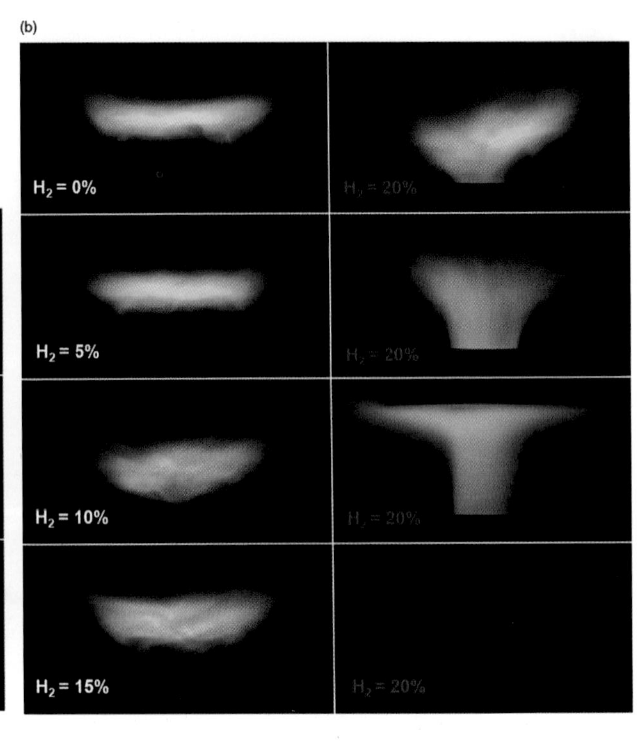

Figure 1.9

1.2.4 Differential Diffusion Effects in Hydrogen-Enriched Laminar Methane Flame at an Equivalence Ratio of 0.8

Ehsan Abbasi-Atibeth and Jeffrey M. Bergthorson

McGill University

The effects of differential diffusion in premixed flames are shown in a hydrogen-enriched methane-air flame at equivalence ratio of 0.8. By increasing the hydrogen content in the fuel, the Lewis number of the fuel and air mixture is reduced, and the mixture becomes more sensitive to differential diffusion effects.

The left images show a laminar flame stretched mostly due to the bulk strain rate. As the hydrogen content is increased, the flames are observed to move upstream, towards the unburned mixture. In a stagnation flame, this is direct visual evidence that the stretched flame speed is increasing, while the unstretched laminar flame speed is held constant. Finally, the flame becomes unstable and close to flashback at 90% hydrogen content.

In the turbulent case (right side), flame stretch is not only due to the bulk stain rate, but also due to the curvature effects of the turbulent eddies in the flow. Therefore, as seen in the image, the turbulent flame is more sensitive to the effects of differential diffusion. It is illustrated that by increasing hydrogen content of the fuel, the flame is burning faster and moves upstream into the unburned reactants, and consequently, flash back occurs at hydrogen content of 20%, which is much lower than in the laminar case.

This research was funded by the Natural Sciences and Engineering Research Council of Canada and Siemens Canada under the Collaborative Research and Development program.

Keywords:
Hydrogen; laminar; turbulent; differential diffusion

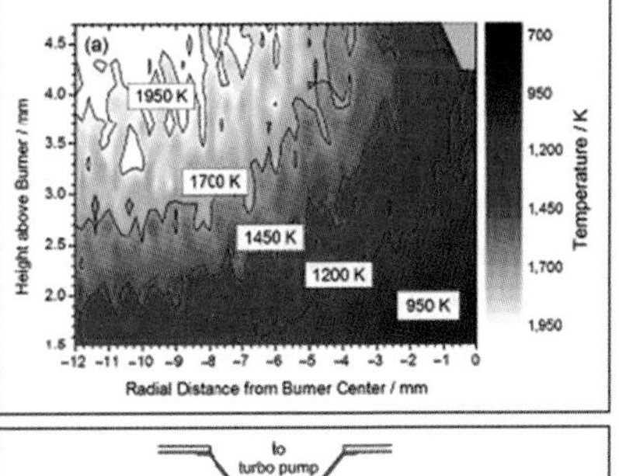

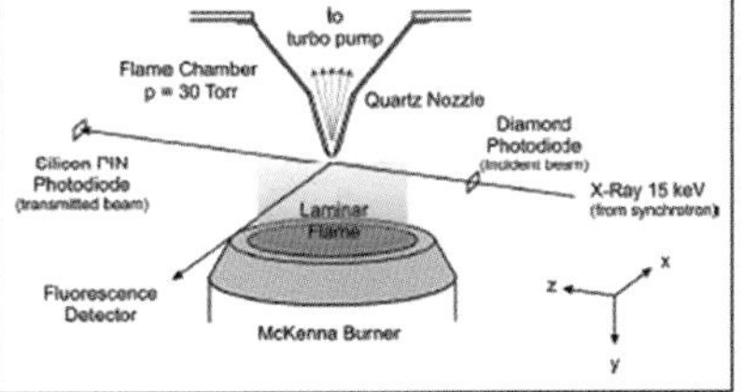

Figure 1.10

1.2.5 Two-Dimensional Imaging of the Temperature Field in Laminar Low-Pressure Premixed Flames

Nils Hansen

Sandia National Laboratories

Robert Tranter and Alan Kastengren

Argonne National Laboratory

The left-side image shows the experimental set-up used at the 7BM beamline of the Advanced Photon Source of the Argonne National Laboratory to measure the two-dimensional temperature field in laminar low-pressure premixed fuel-rich (stoichiometry of 1.7) ethylene (C_2H_4)/O_2/Ar/Kr flame with a cold gas composition by volume of 14.5% C_2H_4, 22.5% O_2, 55% Ar and 5% Kr at a total flow rate of 4 slm. The pressure was kept at 30 Torr. The flame appears in yellow because of the polyimide windows used for the X-ray experiments. The experiments used X-ray florescence spectroscopy to measure the concentration of Kr atoms and hence determine the local temperature in the flame at each measurement location. A highly focused beam (waist 6×7 μm) of 15keV X-rays was used to excite the Kr atoms. The kα florescence photons were collected orthogonal to the X-ray beam by confocal microscopy from a volume (300×6×7 μm) allowing detailed maps of the temperature field to be obtained and the extent to which a probe inserted into the flame distorts the temperature field to be determined.

The smaller images (right side) show a close-up of the burner-stabilized flame, a schematic of the experimental set-up, and some results of measurements of the temperature around a sampling probe.

This material is based upon work supported by the U.S. Department of Energy (DOE), Office of Science, Office of Basic Energy Sciences.

Keywords:
Premixed; low pressure

Reference

N. Hansen, R. S. Tranter, K. Moshammer, J. B. Randazzo, J. P. A. Lockhart, P. G. Fugazzi, T. Tao, A. L. Kastengren, 2D-Imaging of Sampling-Probe Perturbations in Laminar Premixed Flames using Kr X-ray Fluorescence, Combustion and Flame, 181, pp. 214–224, 2017.

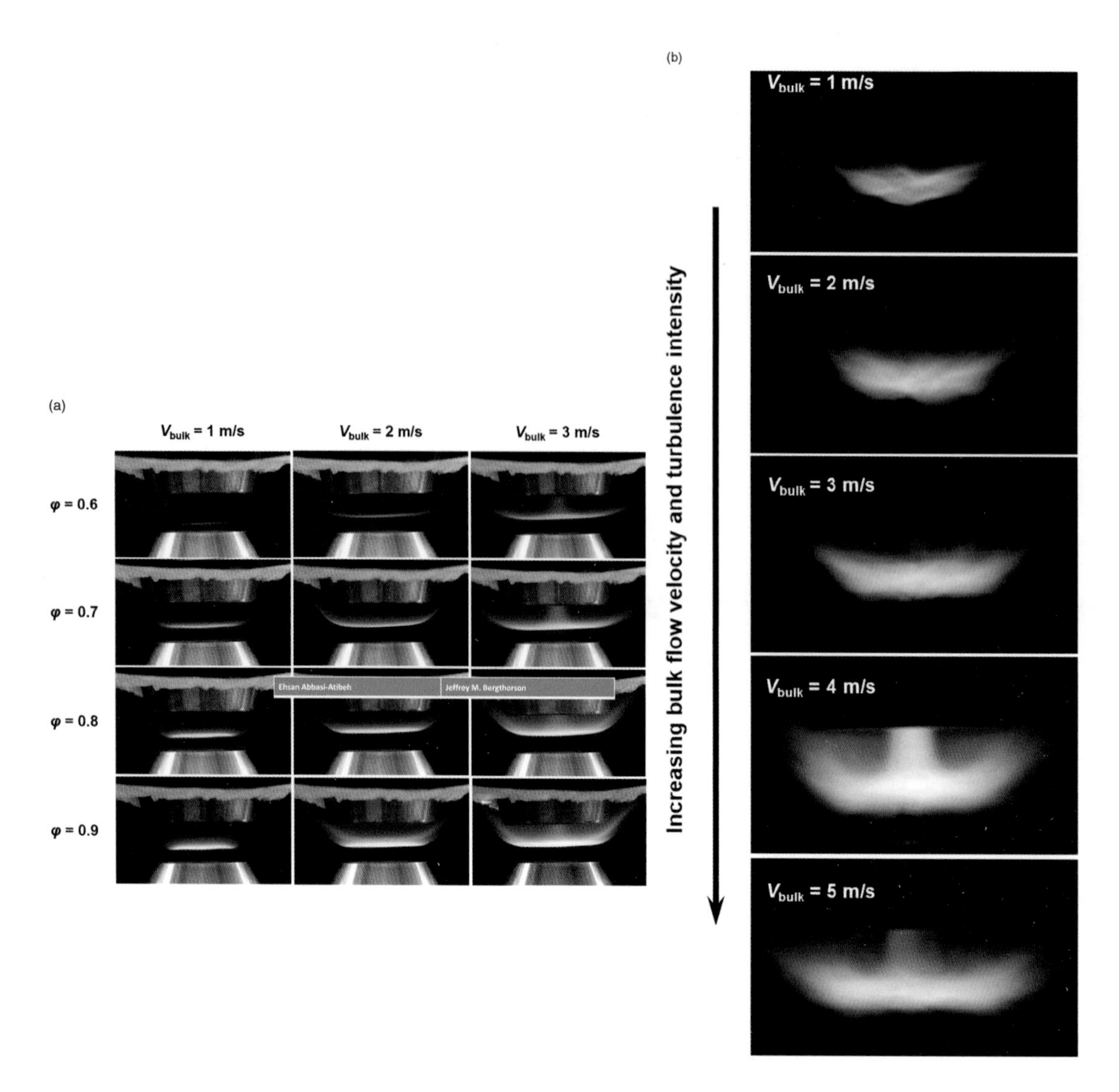

Figure 1.11

1.2.6 Aerodynamically Stabilized Laminar Flames at Various Equivalence Ratios and Bulk Strain Rates

Ehsan Abbasi-Atibeth and
Jeffrey M. Bergthorson
McGill University

Lean premixed flames are stabilized in the Hot exhaust Counter-flow Turbulent flame Rig (HCTR) at various equivalence ratios and turbulence intensities. The rig is designed to stabilize laminar and turbulent flames against a hot product flow in an axial opposed-flow configuration. A co-flow of inert gas is used to reduce the effect of the shear layer that is formed between the inner flows and the ambient air and shrouds the opposed flow from surrounding air.

In left images, laminar flames are stabilized against hot products at equivalence ratios ranging from 0.6 to 0.9, and the bulk flow velocity at the premixed fuel and air nozzle exit ranges from 1m/s to 3 m/s. Right images show stabilized turbulent flames against hot products at various bulk flow velocities and turbulence intensities ranging from weakly wrinkling flames to turbulent flames in the thin reaction zone regime (top to bottom). Increasing the bulk flow velocity at the premixed fuel and air nozzle exit, which ranges from 1m/s to 5m/s, increases turbulence intensity in these tests.

This research was funded by the Natural Sciences and Engineering Research Council of Canada and Siemens Canada under the Collaborative Research and Development program.

Keywords:
Aerodynamically stabilized; bulk strain rate

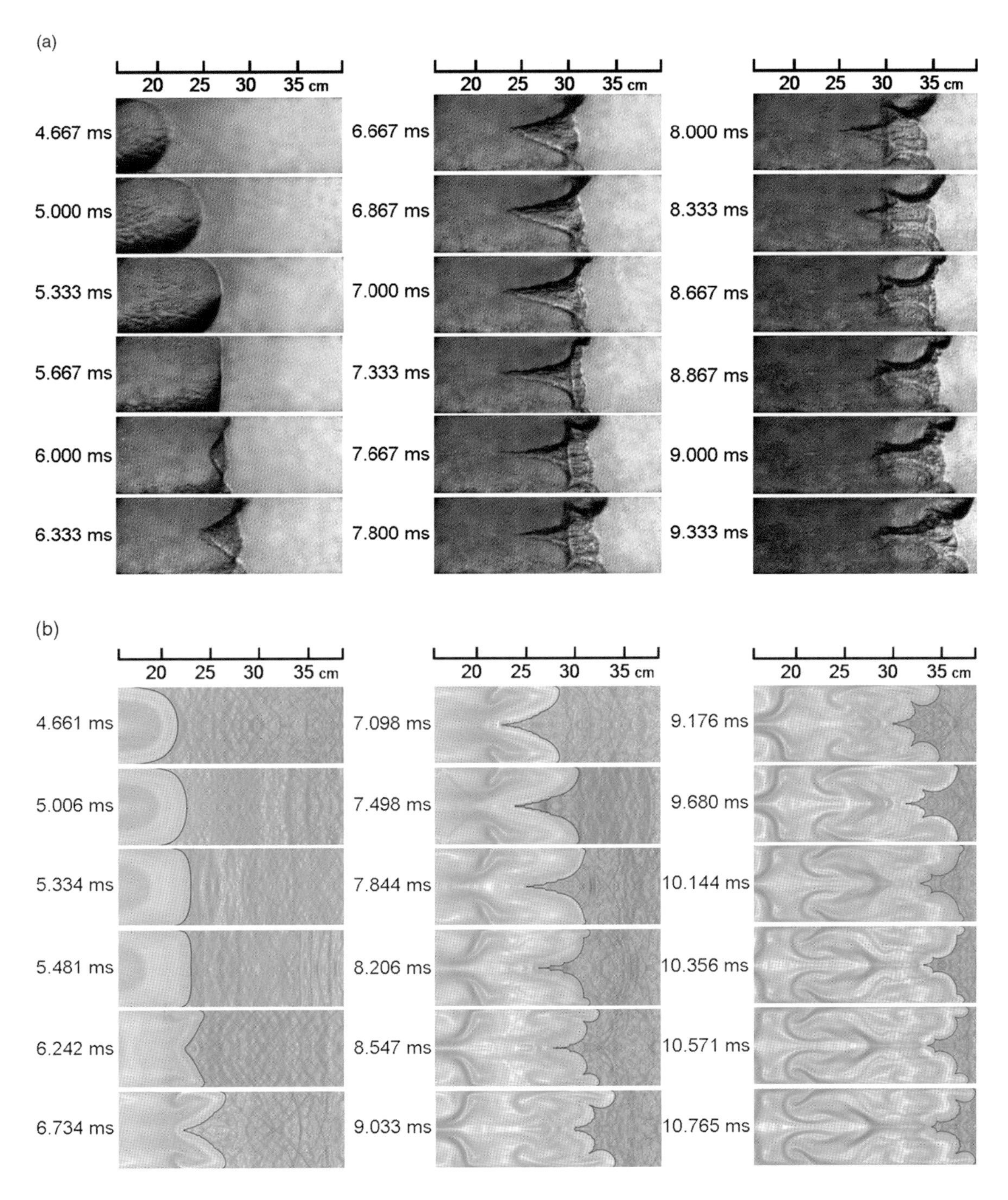

Figure 1.12

1.2.7 Tulip Flame and Distorted Tulip Flames in Experiments and Numerical Simulations

Michael Gollner

University of California, Berkeley

The first row shows a sequence of schlieren images illustrating the development of tulip flame and distorted tulip flames in experiments. The second row shows schlieren images taken from numerical simulations. Both the experiment and the simulation show that the flame shape first changes from convex to concave, and then to a tulip flame, and then distorted tulip flames. The distorted tulip flames develops into a triple tulip flame as the secondary cusps approach the center of the primary tulip flame lips (at 7.8 ms in experiments and 9.033 ms in simulations). A second distorted tulip flame is generated with a cascade of distortions superimposed on the primary lips (at 9.0 ms in experiments and 10.571 ms in simulations) before the disappearance of the first distorted tulip flame.

Keywords:

Tulip flame; experimental; numerical

Figure 1.13

1.2.8 Transition in a Turbulent Flame

Bret Windom, Bo Jiang, Sang Hee Won, and Yiguang Ju

Princeton University

The series of eleven n-heptane/air flames demonstrate the transition that occurs in a turbulent flame as a result of low-temperature oxidation of the reactants prior to their introduction into the high-temperature flame. Scanning left to right, the degree of pre-flame reactant oxidation is increased by increasing the reactant temperature and/or the heated residence time. This transition, evident by the blossoming redness of the flames, can have serious implications on the flame properties, including burning rates, emissions, turbulent/combustion interactions, and flame regimes.

Keywords:

Heptane/air; turbulent

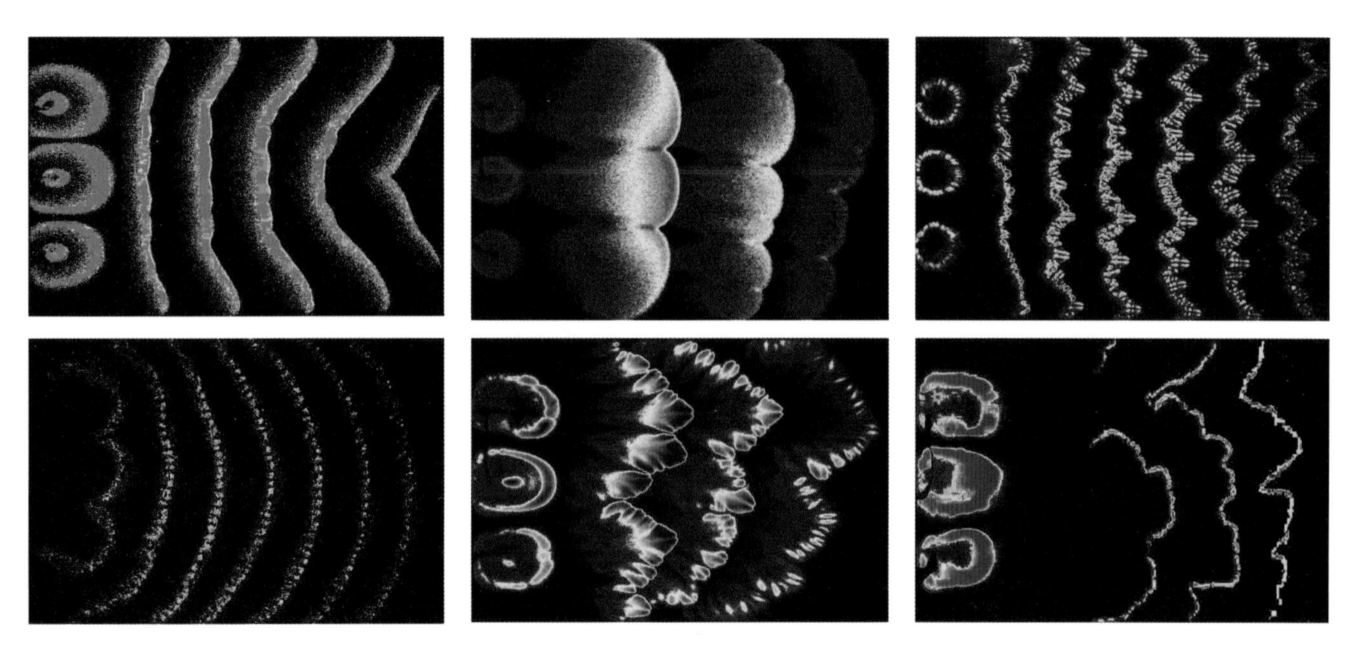

Figure 1.14

1.2.9 Flame Propagation in Narrow Channels at Varying Lewis Number

Si Shen

University of California, Los Angeles

A Hele-Shaw cell is used for this laminar flame instability experiment. Dimensions of entire visible area are 40.0 cm by 60 cm. The Hele-Shaw cell is first evacuated, and then filled with desired premixed gas mixture. The gas mixture is ignited at one end of the cell by three ignition sparks. The flame propagates under constant pressure toward the closed end of the cell with ignition end open to the atmosphere. There is no turbulence pre-ignition in the cell. Light emitted from hot products behind flames is filmed in total darkness at 100 frames per second. The photo is created by collapsing frames of the video. Each frame is separated by equal time intervals. The photo is also colorized base on temperature gradient with low temperature colored in blue, and high temperature colored in green.

Top left to right images: (1) H2-O2-N2 flames at equivalence ratio, $\phi = 2$, calculated adiabatic flame temperature, Ta = 1200K, downward propagating, gap between plates, d = 1.27 cm; (2) H2-O2-N2 flames at $\phi = 2$, Ta = 1200K, horizontally propagating, d = 1.27 cm; (3) H2-O2-CO2 flames at $\phi = 0.35$, Ta = 1050K, downward propagating, d = 1.27 cm.

Bottom left to right horizontally propagating flame image: (1) H2-O2-N2 flames at $\phi = 0.45$, Ta = 900K, d = 1.27 cm; (2) H2-O2-N2 flames at $\phi = 0.8$, Ta = 1300K, and d = 1.27 cm; (3) H2-O2-N2 flames at $\phi = 0.8$, Ta = 1300K, and d = 0.32 cm.

This research was supported by the National Science Foundation under Grant CBET-1236892.

Keywords:
Flame propagation; Lewis number

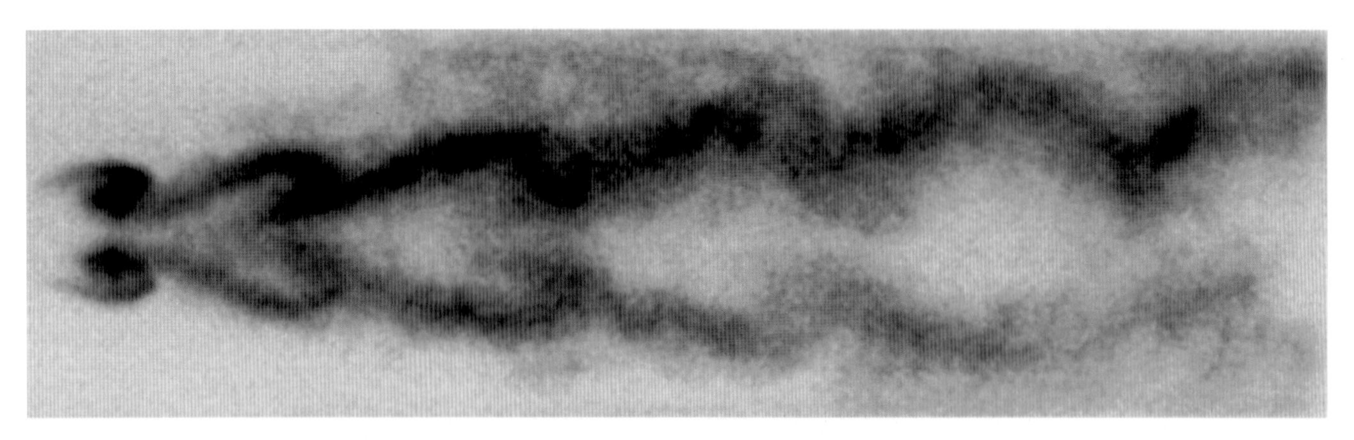

Figure 1.15

1.3 Practical Flames

1.3.1 Chemiluminescence Image of a Longitudinally Excited Bluff Body Flame Wrinkled by Coherent Vortical Wake Structures

Benjamin Emerson and Tim Lieuwen

Georgia Tech University

This is a chemiluminescence image of an acoustically forced bluff body flame. The exposure time of the image is 200 microseconds. The image shows a wrinkled flame structure that was produced by varicose (symmetrically shed) vortex structures. This is a demonstration of the velocity-coupled flame response, where acoustic oscillations excite hydrodynamic instabilities, which amplify vortical structures and consequently wrinkle the flame. The combustor was supplied with non-vitiated, non-preheated air. The combustor was operated at atmospheric pressure, at a velocity of 50 m/s and a premixed air/natural gas equivalence ratio of 0.8. The combustor was longitudinally acoustically forced with planar acoustic waves at 515 Hz.

Keywords:

Chemiluminescence; bluff body; coherent vortical wake structures

Reference

Emerson, B. and Lieuwen, T., 2015. Dynamics of harmonically excited, reacting bluff body wakes near the global hydrodynamic stability boundary. Journal of Fluid Mechanics, 779, pp.716–750.

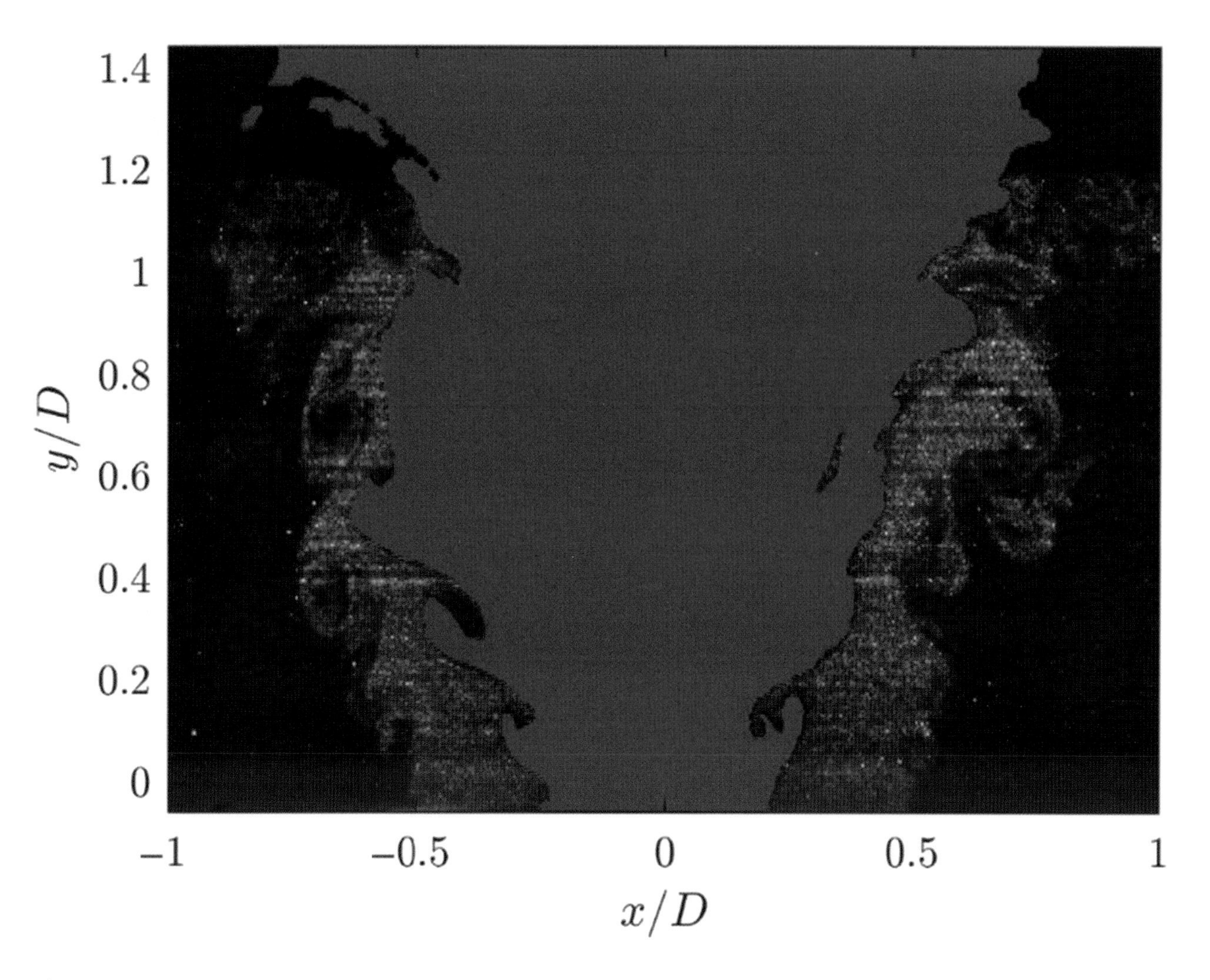

Figure 1.16

1.3.2 Overlay of a Filtered Instantaneous OH-PLIF (Blue) Image and Raw Mie Scattering (Yellow) for a Premixed Methane–Air Swirl Flame

Travis E. Smith, Ianko Chterev, Benjamin L. Emerson, and Timothy C. Lieuwen

Georgia Tech University

This image is a composite of two false-colored images. The yellow color belongs to planar Mie scattering from alumina particles with which the flow was seeded for flow visualization and a stereoscopic particle image velocimetry measurement. The blue color belongs to planar laser induced fluorescence (PLIF) of OH. The Mie scattering data provide a visualization of the vortex structures that dominate the fluid dynamics, and the boundaries of the OH PLIF measurement indicate the premixed flame front position. The two images were measured simultaneously and capture a measurement plane that intersects the combustor centerline. This composite demonstrates the fluid-dynamic wrinkling of the flame that often governs the flame dynamics in swirl stabilized combustors. These images were acquired in a lean premixed combustor operating at atmospheric pressure and a preheat temperature of 450 K.

This research was partially supported by the Air Force Office of Scientific Research under award no. FA9550-16-1-0442.

Keywords:

Methane; premixed; swirl; Mie scattering

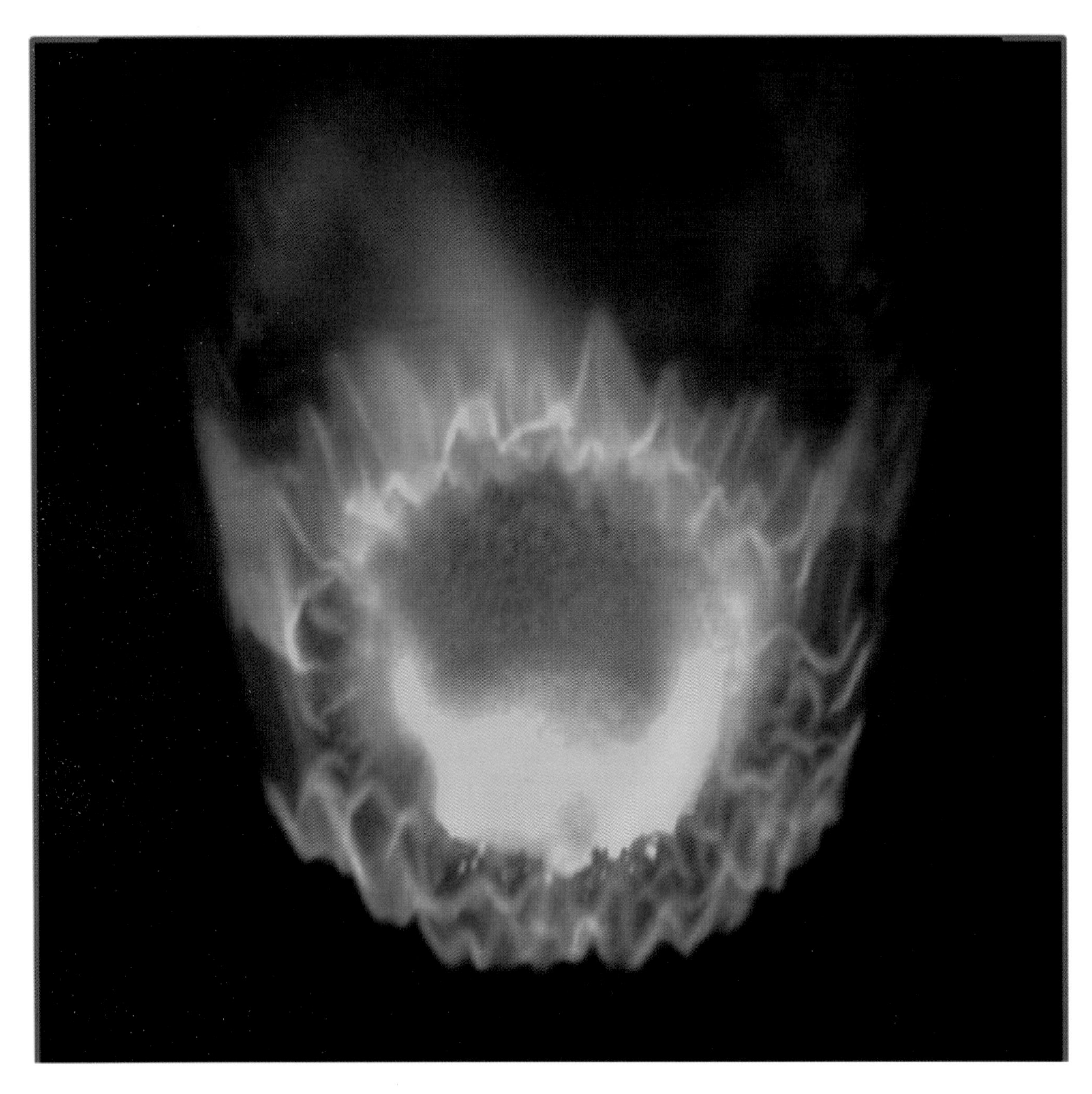

Figure 1.17

1.3.3 Porous Inert Media with Stable Methane Flame

L. Justin Williams and Ajay K. Agrawal
University of Alabama

This is a lean premixed flame stabilized by two different techniques. The bulk flame in the center is a swirl-stabilized flame and it is surrounded by a ring of flamelets stabilized on the downstream surface of an annular porous insert. Premixed reactants enter the combustor, and the mass flow split between the bulk flame and flamelets is determined by the porous insert geometry. Flamelets provide thermal feedback to the bulk flame to enhance static flame stability without the outer flow recirculation region. Porous insert minimizes interactions among vortical structures and reactions zones to mitigate thermoacoustic instabilities that can occur in such systems without the insert. The flame produces a sound pressure level of 92.1 dB at an equivalence ratio of 0.7, which is about 20 dB lower than the sound produced by a similar flame without the porous insert.

Keywords:
Methane; porous inert media

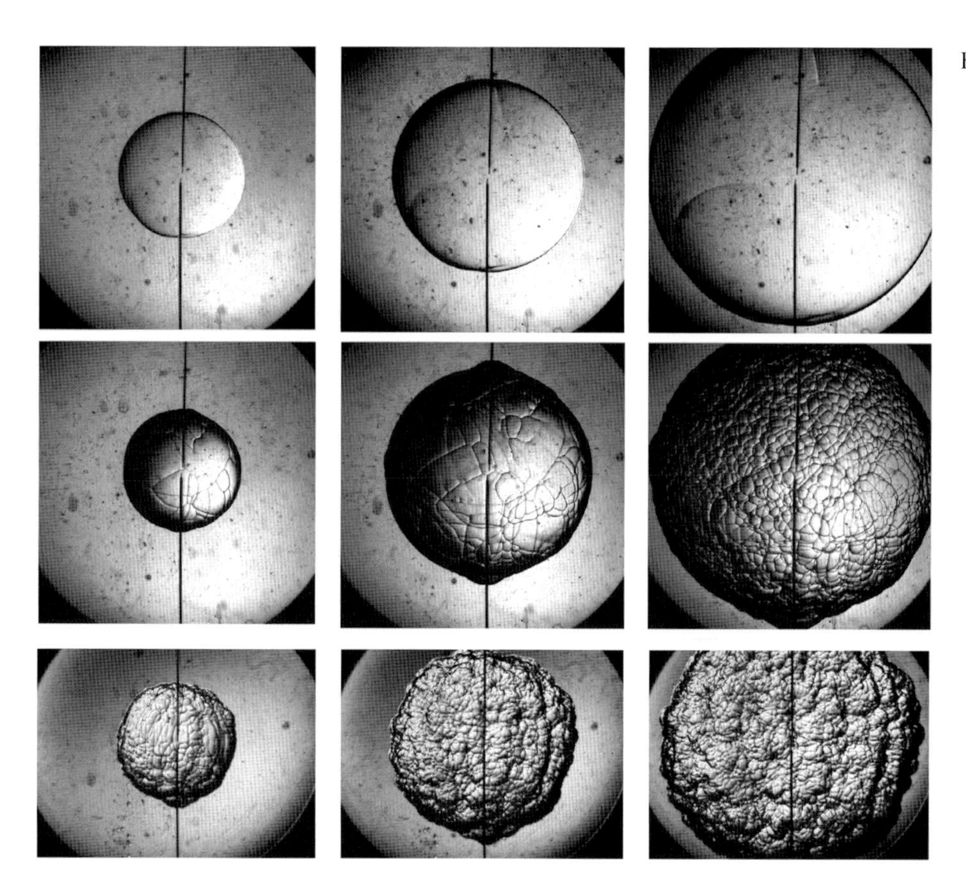

Figure 1.18

1.4 Spherical Flames

1.4.1 Cellular Instabilities in Outwardly Propagating Flame

Abhishek Saha and Chung K. Law
Princeton University

Two instability mechanisms are intrinsic to the propagation of laminar premixed flames, namely the hydrodynamics, Darrieus-Landau (DL), instability and the diffusional-thermal (DT) instability. DL instability considers the flame to be infinitely thin and structureless, locally propagating with the (constant) laminar flame speed; cells develop due to the local imbalance in the flame and flow speeds. Consequently, DL instability is either aggravated or moderated with decreasing or increasing flame thickness, respectively, with a wide range of cell sizes exceeding the flame thickness. On the other hand, DT instability considers the imbalance in the thermal and reactant mass diffusivities, which is characterized by their ratio, the mixture's Lewis number, Le, such that the flame is DT unstable and stable for $Le<1$ and >1, respectively, while the cell sizes are of the order of the flame thickness and as such are smaller than the DL cells.

Based on the controlling mechanisms described above, DL instability is promoted with increasing system pressure, which reduces the flame thickness, and is relevant to processes within internal combustion engines, while DT instability is promoted if the controlling, deficit reactant is also a light species, as for lean hydrogen/air or rich heavy-hydrocarbon/air mixtures. These features are demonstrated here in the morphologies of spark-ignited outwardly propagating flames in a constant-pressure chamber.

Row 1: the flame sequence does not exhibit any flame wrinkling because the pressure (1 atm) is not high (no DL instability) while the deficit reactant, O_2, is also a heavy molecule ($Le>1$).

Row 2: the flame sequence exhibits DL instability because of the elevated pressure (5 atm.) but not the DT instability ($Le>1$).

Row 3: the flame sequence does not exhibit DL instability (pressure at 2 atm) but exhibit DT instability because the deficit reactant, H_2, is a light molecule ($Le<1$).

This research was supported by the National Science Foundation.

Keywords:
Cellular instabilities

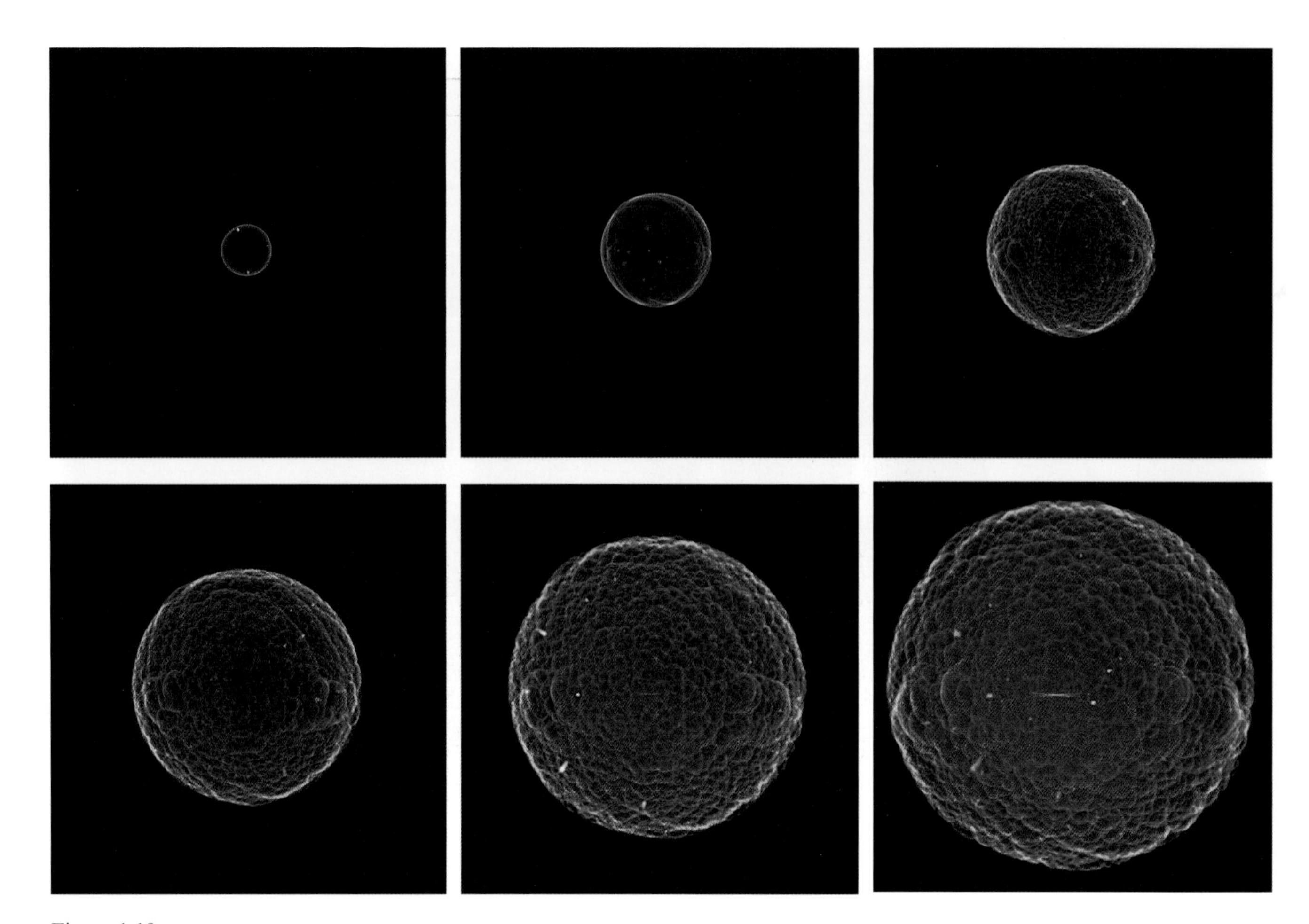

Figure 1.19

1.4.2 Growth of Darrieus–Landau Instability in a Large-Scale Spherical Flame

C. Regis Bauwens, Jeffrey M. Bergthorson, and Sergey B. Dorofeev

FM Global and McGill University

An image sequence of the growth of flame instabilities on a premixed stoichiometric propane-air flame is shown. This experiment was performed under quiescent initial conditions at standard temperature and pressure in a 64 m^3 enclosure vented to maintain a constant pressure. The image sequence shows the growing spherical flame at diameters of 0.2, 0.4, 0.6, 0.8, 1.0, and 1.2 m.

Initially, the flame is stabilized by stretch due to the increase in flame radius, reducing the amplitude of any initial disturbance created at the time of ignition. At a radius of 0.2 m, however, the growth of large-scale flame cracks is clearly seen as the flame grows from 0.2 – 0.4 m diameter. At this point, the onset and growth of the Darrieus-Landau instability occurs, where the spontaneous formation of cellular structures appear across the entire flame surface.

This work was funded by FM Global and was performed within the framework of the FM Global Strategic Research Program on Explosions and Material Reactivity.

Keywords:

Darrius–Landau instability; spherical flame

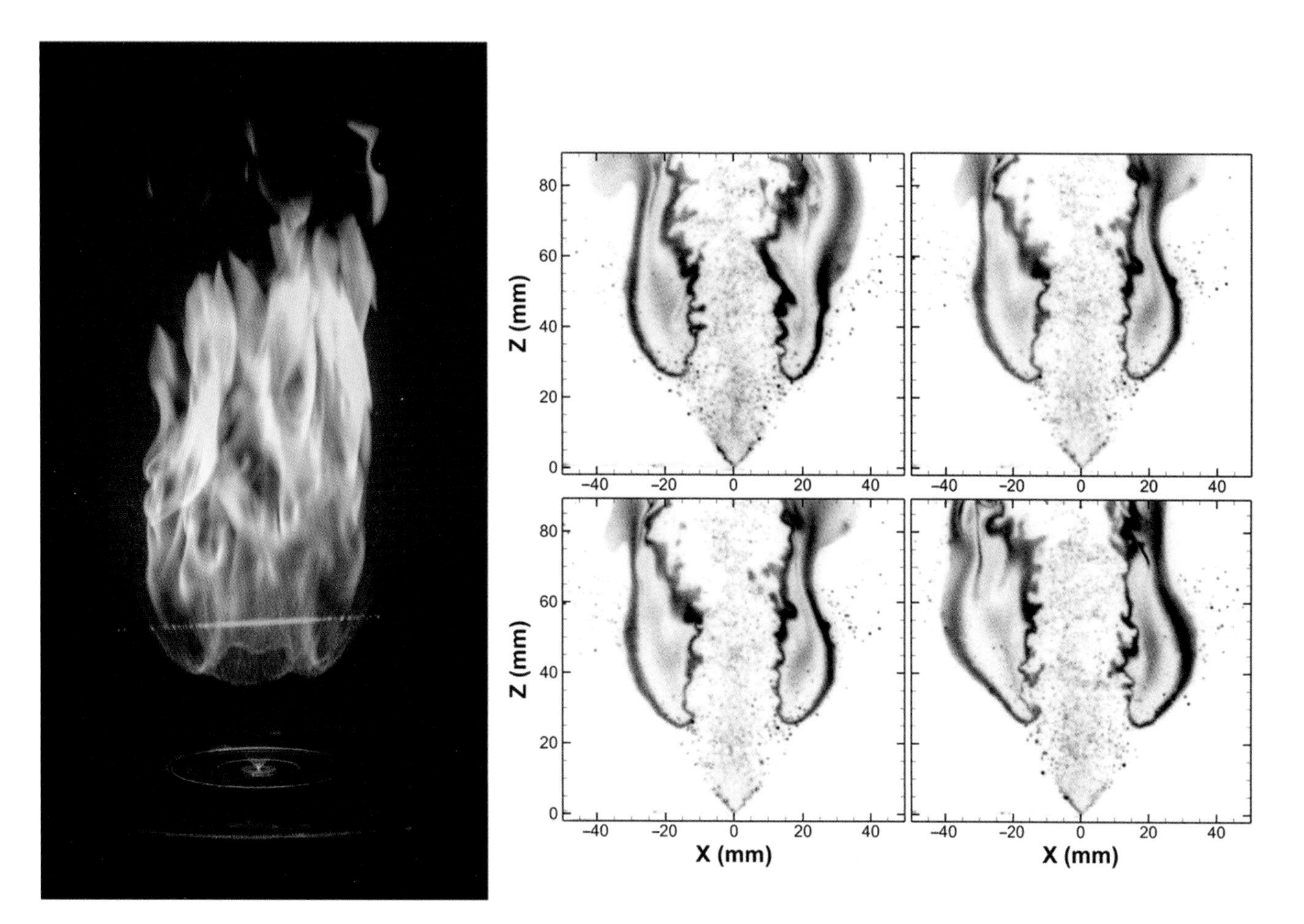

Figure 1.20

1.5 Gas Jet and Liquid Flames

1.5.1 Complex Nature of Jet Spray Flame Structure

Antoine Verdier, Javier Marrero Santiago, Gilles Cabot, and Bruno Renou
CORIA, INSA

Left image shows a lifted n-heptane jet spray flame. The fuel injection system is composed of a simplex fuel injector (Danfoss, 1.35 kg/h, 80 degrees, hollow cone) and an external annular, non-swirling air co-flow with an inner and outer diameter of 10 and 20 mm respectively. The diameter of the injector orifice is 200 microns. A continuous laser beam (532 nm) is crossing the flame revealing the fuel droplet density by Mie scattering.

Right side images show instantaneous snapshots of OH-PLIF at 10 Hz in a vertical cross section and reveals the complexity of jet spray flame structure, including different reactive zones: a wrinkled inner reactive zone (IRZ) and a non-premixed outer diffusion zone (ORZ), connected by the flame leading edge. The field of view is $112 \times 112 mm^2$ leading to a magnification ratio of 0.2212 mm/pix.

The authors gratefully acknowledge the financial support from the Agence Nationale de la Recherche (ANR) under the project TIMBER ANR-14-CE23-0009.

Keyword:
Jet spray flame

Reference

Verdier, A., Marrero Santiago, J., Vandel, A., Saengkaew, S., Cabot, G., Grehan, G., & Renou, B. (2017). Experimental study of local flame structures and fuel droplet properties of a spray jet flame. Proceedings of the Combustion Institute, 36(2), 2595–2602.

Figure 1.21

1.5.2 Reacting Fuel Spray from a Diesel Injector

Joshua Bittle, Ajay K. Agrawal, Christopher Wanstall, and Ross Depperschmidt

University of Alabama

Reacting n-heptane spray flame acquired by direct photography. The fuel is injected (from left side) at 100 MPa for 3ms through a customized diesel injector which has a single 100 μm diameter axial hole. The ambient is a continuous flow of air at 3 MPa and 800 K. The image shown was acquired immediately after the end of injection at 3 ms. The injected fuel autoignites approximately 1.8 ms after the start of injection. For scale referencing, a 3 mm diameter thermocouple probe can be seen mounted at the top near the injector inlet.

Keywords:

Diesel injector; fuel spray

Figure 1.22

1.5.3 Boilover of a Crude Oil Pool Fire Burning on Water

Laurens van Gelderen
Technical University of Denmark

Grunde Jomaas
University of Edinburgh

Boilover of a crude oil pool fire on water (contained in a Pyrex glass cylinder with a diameter of 0.16 m). The water inside the Pyrex glass cylinder is trapped below the burning oil (top left), thus preventing the water from vaporizing as it gets heated up. This causes the water to superheat to a temperature of 120 to 150 °C, at which point the water violently evaporates through the oil slick, ejecting steam and oil droplets into the flame (bottom left). The explosive nature of steam and oil droplets inside the flame cause a chain reaction that drastically increases the burning rate and flame height of the oil, known as boilover (right).

The project was funded by the Danish Council for Independent Research Grant DDF – 1335-00282.

Keywords:
Crude oil; pool fire

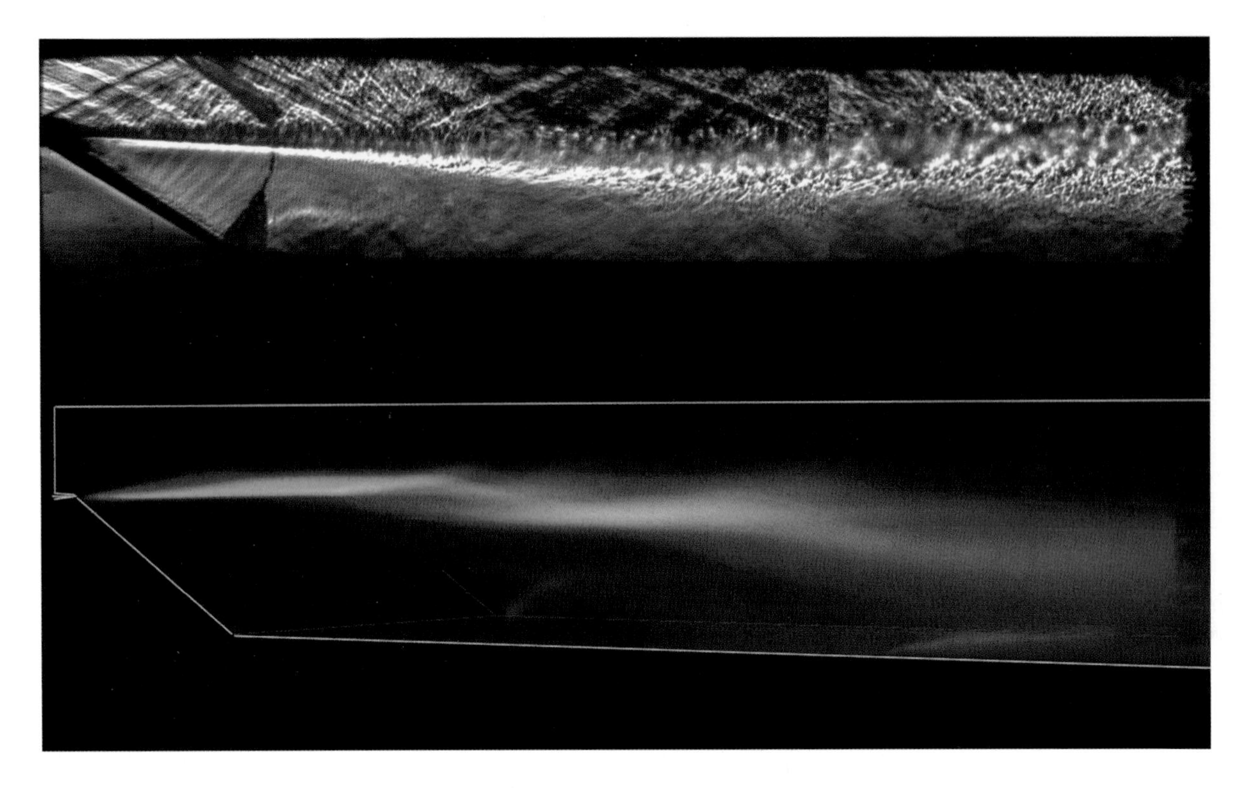

Figure 1.23

1.6 Flames at High Speed

1.6.1 Combustion of Hydrogen and Fluorine in a Supersonic Shear Layer at Mach 2.5

Jeffrey M. Bergthorson, Aristides M. Bonanos, and Paul E. Dimotakis

California Institute of Technology

The top image is a composite schlieren image stitched together from three experimental runs. The bottom image shows a visible image of the chemiluminescence that results from the reacting hydrogen-fluorine shear layer under the same flow and combustion conditions. The chemiluminescence image is taken with a perspective looking upstream, and the test chamber geometry is indicated by the white lines.

The top-stream flow has a Mach number of 2.5, corresponding to a velocity of 570 m/s, and a composition of 19.75% hydrogen and 0.5% nitric oxide in a bulk of nitrogen. The secondary bottom-stream flow issues from a perforated plate, aligned at an angle at the lower left of the images, at a speed of 60 m/s. The bottom-stream flow contains 5% fluorine in a bulk of nitrogen. The reaction rate between fluorine and hydrogen, catalyzed by nitric oxide, is much faster than the characteristic mixing rates, enabling the reaction to be used to measure molecular mixing rates.

The reflecting expansion and shock waves, characteristic of the supersonic Mach 2.5 flow, are visible in the top stream in the schlieren image. The growth of the compressible turbulent boundary layer along the top wall of the facility can be noticed. The reacting shear layer, between the hydrogen and fluorine, is visible in both images. In the composite Schlieren image, the peak temperature location in the shear layer is indicated by the transition from relatively dark to bright intensity. This effect is more distinct at the left of the image since the strong turbulence reduces the temperature gradients in the downstream direction. A rake of pitot probes and thermocouples are visible at the right-hand side of each image.

This work was funded by the AFOSR under Grants FA9550-04-1-0020 and FA9550-04-1-0389.

Keywords:
Hydrogen/fluorine; supersonic; shear layer

Reference

AM Bonanos, JM Bergthorson, and PE Dimotakis. Molecular mixing and flowfield measurements in a recirculating shear flow. Part II: Supersonic flow. Flow, Turbulence & Combustion 83 (2009), 251–268.

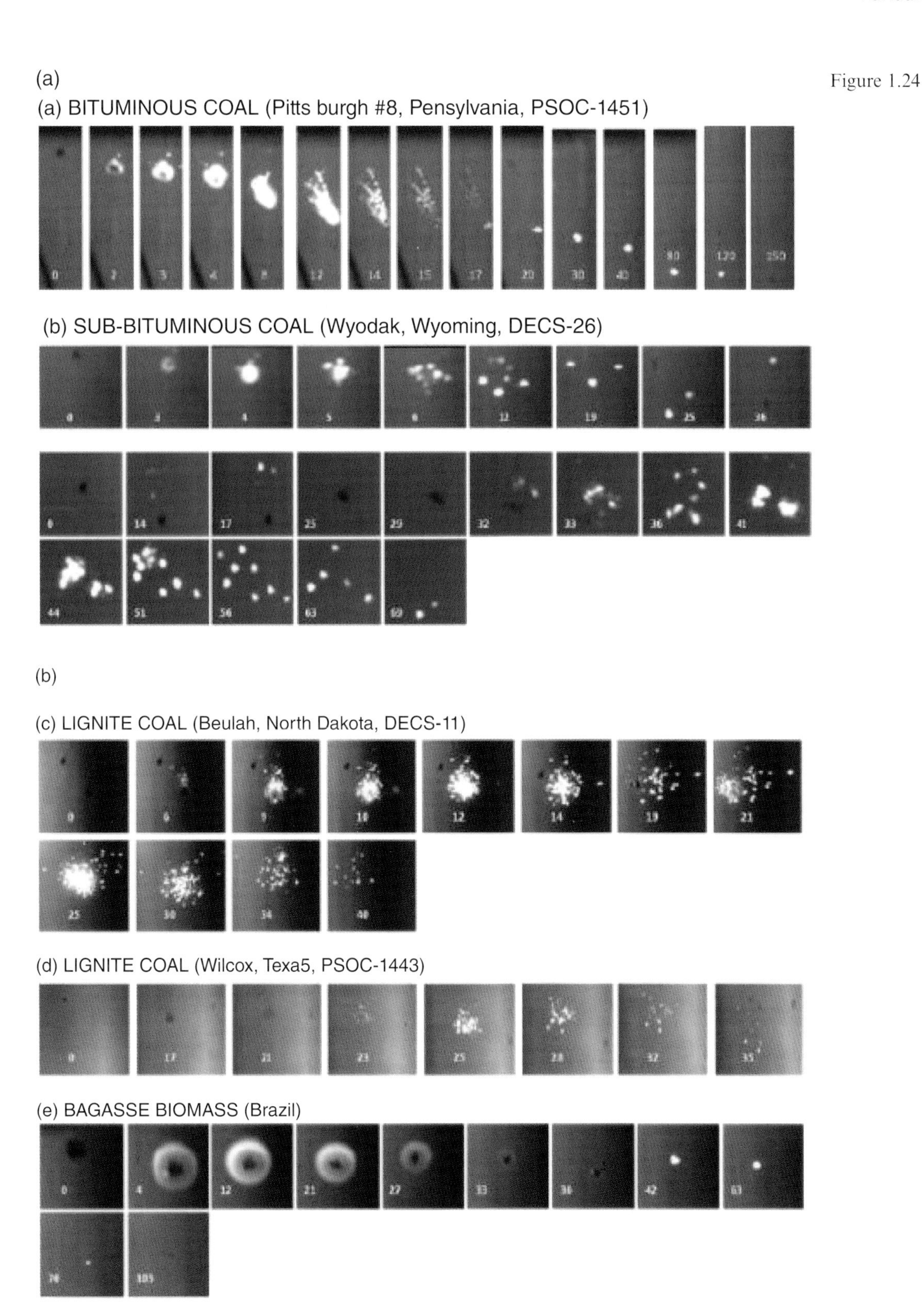

Figure 1.24

1.7 Coal and Solid Particle Flames

1.7.1 Combustion Behavior of Single Particles from Three Different Coal Ranks and from Biomass

Kulbhushan A. Joshi and Yiannis A. Levendis
Northeastern University

Selected images from high-speed cinematography of typical combustion events involving particles of five fuels (bituminous, sub-bituminous, two lignite coals and sugarcane bagasse biomass) free-falling in an electrically-heated laminar-flow transparent drop-tube furnace. Particle sizes were in the range of 74–90 μm and the furnace was operated under atmospheric conditions and a wall temperature set to 1400 K (Tgas approx. 1350 K). The diameter of the wire shown in sequence (a) is

250 μm. The displayed numbers in each frame are in milliseconds, where zero represents beginning of the depicted sequence.

In the case of the sub-bituminous coal two sequences are included, one where volatile flame was present (top), and one where it was absent (bottom). In the depicted frames of the Beulah lignite two particles ignite and burn in similar fashion. (a) bituminous coal (Pittsburgh #8, Pennsylvania, PSOC-1451), (b) sub-bituminous coal (Wyodak, Wyoming, DECS-26), (c) lignite coal (Beulah, North Dakota, DECS-11), (d) lignite coal (Wilcox, Texas, PSOC-1443), and (e) sugarcane bagasse biomass (Brazil).

The authors acknowledge financial assistance from the NSF award CBET-0755431.

Keywords:
Coal; biomass; single particle behavior

Reference

Levendis, Y.A., Joshi, K., Khatami, R. and Sarofim, A.F., 2011. Combustion behavior in air of single particles from three different coal ranks and from sugarcane bagasse. Combustion and Flame, 158(3), pp. 452–465.

Figure 1.25

1.7.2 Coal Dust Passing through a Premixed Methane/Air Flame

Scott R. Rockwell

University of North Carolina, Charlotte

Image of a premixed turbulent methane air flame with coal particles passing through on a stabilized burner is shown. The coal particles are 75-125 microns in diameter with a concentration less than 25 g/m3. The images were taken after tests conducted at coal dust feed rates of 25 g/m3 and what is burning in this picture are the residual coal particles continuing to flow through the feed system. At higher feed rates the premixed methane air flame is not visible through the bright illumination from the coal particles.

Keywords:

Methane/air; coal dust; premixed

Reference

Rockwell, Scott R., Rangwala, Ali, S., Influence of Coal Dust on Premixed Turbulent Methane-Air Flames, Combustion and Flame 2013, Vol. 160, (3) 635–640.

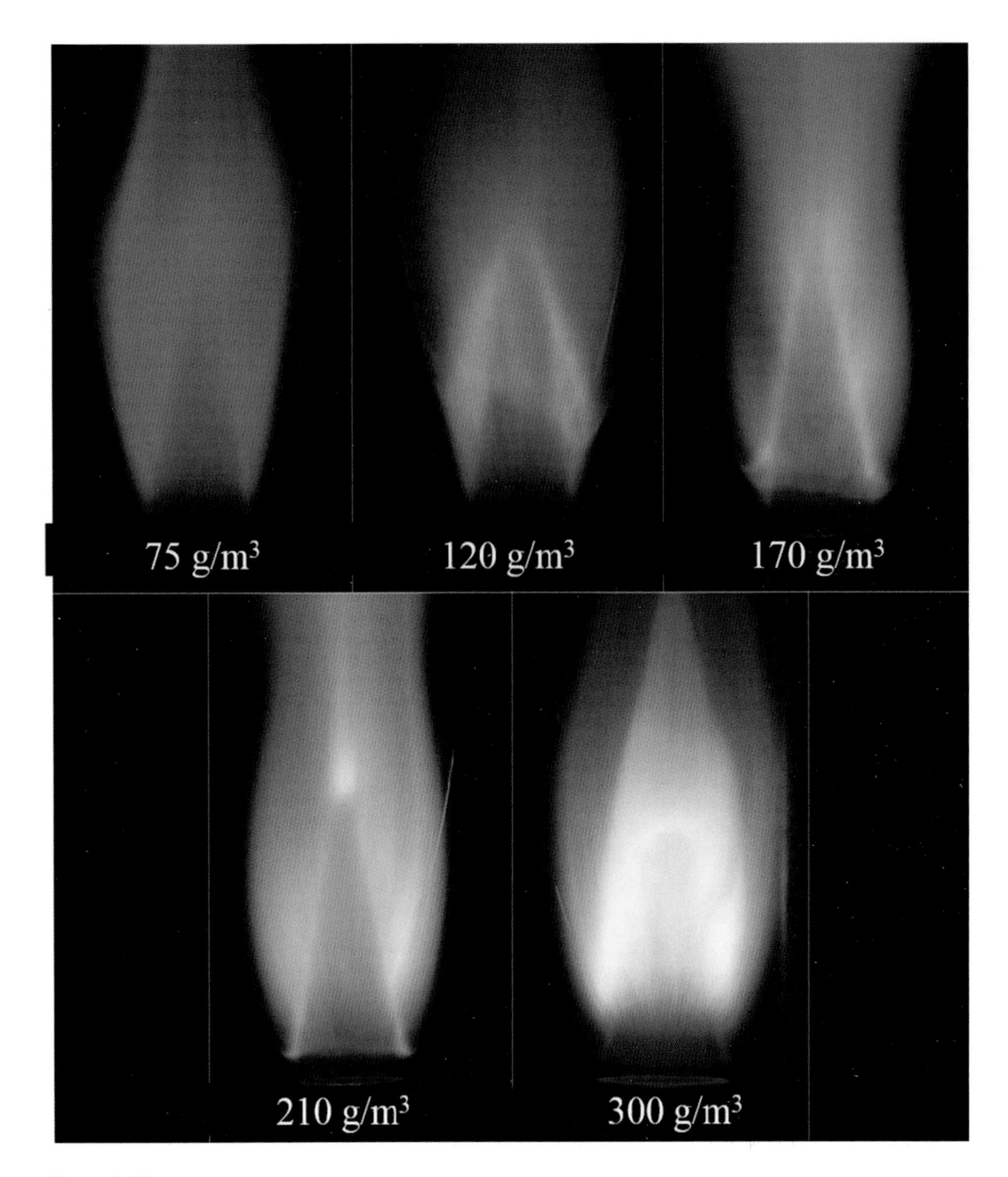

Figure 1.26

1.7.3 Aluminum–Methane/air Flame Coupling

Michael Soo, Phillippe Julien, Samuel Goroshin, Jeffrey Bergthorson, David Frost, and Andrew Higgins

McGill University

Bunsen burner flames are shown of aluminum suspensions (Sauter-mean diameter of 5.6 μm) dispersed in stoichiometric methane/air (21% O2/79% N2) gas mixtures and stabilized at different concentrations of aluminum powder. The figure shows the change in flame structure as the aluminum concentration is raised. At low mass concentrations, below 100 g/m3, the flame appears orange in color and looks the same as a flame seeded with inert SiC powder. This indicates that the aluminum does not ignite at low concentrations. As the concentration of aluminum particles increases to roughly 120–140 g/m3, a bright white spot emerges first at the tip of the flame and, as the concentration increases further to about 180 g/m3, a very bright front with a well-defined outer border indicating aluminum combustion moves down along the flame cone, eventually stabilizing just a few millimeters above the rim of the methane flame. The bright white color of the flame results from the high temperatures (~3500K) of the aluminum flame and chemiluminescent emissions from aluminum oxide.

This research was supported by Natural Sciences and Engineering Research Council of Canada and Martec, Ltd., and Defense Threat Reduction Agency under contract HDTRA 1-11-1-0014.

Keywords:
Aluminum–methane/air; coupling

References

Soo M, Julien P, Goroshin S, Bergthorson JM, Frost DL. Stabilized flames in hybrid aluminum–methane–air mixtures. Proc Combust Inst 2013;34:2213–20. http://dx.doi.org/10.1016/j.proci.2012.05.044.

Julien P, Soo M, Goroshin S, Frost DL, Bergthorson JM, Glumac N, et al. Combustion of aluminum suspensions in hydrocarbon flame products. J Propuls Power 2013;30:1–8. http://dx.doi.org/10.2514/1.B35061.

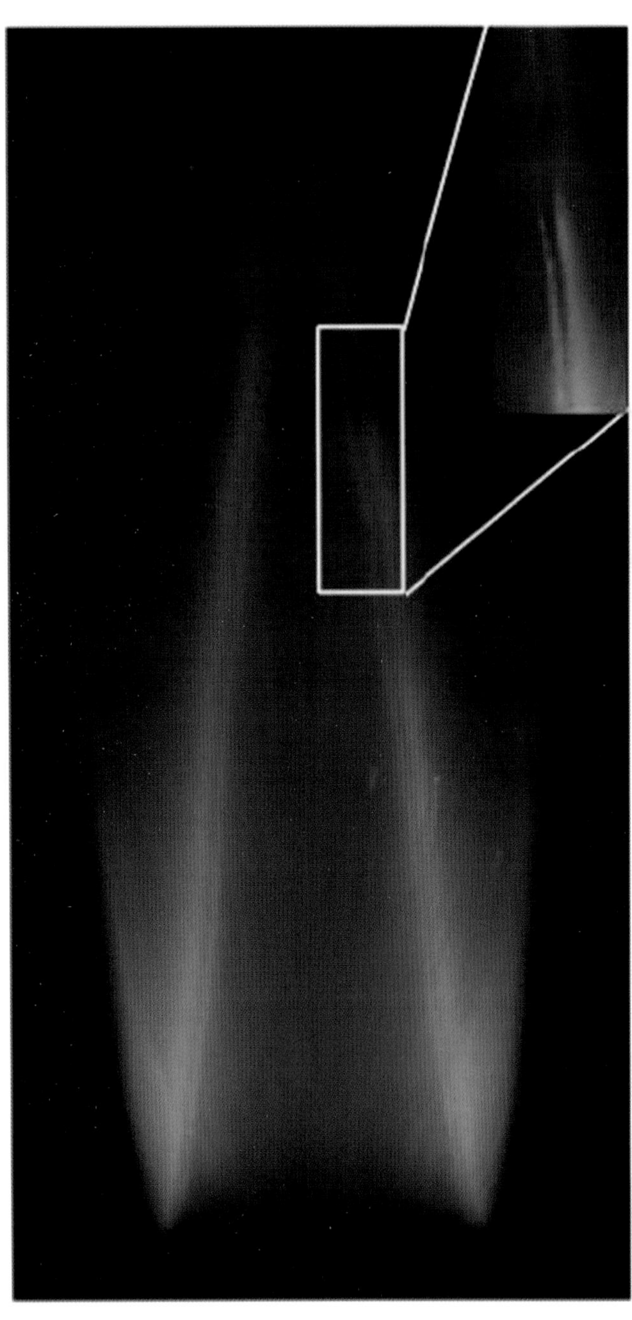

Figure 1.27

1.7.4 Double Front Structure of Methane–Iron/Air Flames

Philippe Julien, James Vickery, Michael Soo, David Frost, Jeffrey Bergthorson, and Samuel Goroshin

McGill University

The photograph shows a stabilized Bunsen iron-methane-air flame. The picture illustrates the double front structure obtained in hybrid metal-hydrocarbon systems. The first front is the methane-air flame. The second front, coupled to the first one, is the iron front, burning in the products of combustion of the methane flame (water and carbon dioxide) at around 2200 K. The methane-air mixture has an equivalence ratio of 1, with no oxygen available for iron combustion, forcing it to burn with steam and carbon dioxide. The iron concentration of the particles (d32 = 2.20 μm) is around 160 g/m3.

Keywords:

Iron–methane/air; double front

Reference

Julien P, Whiteley S, Goroshin S, Soo MJ, Frost DL, Bergthorson JM. Flame structure and particle-combustion regimes in premixed methane–iron–air suspensions. Proc Combust Inst 2015;35:2431–8. http://dx.doi.org/10.1016/j. proci.2014.05.003.

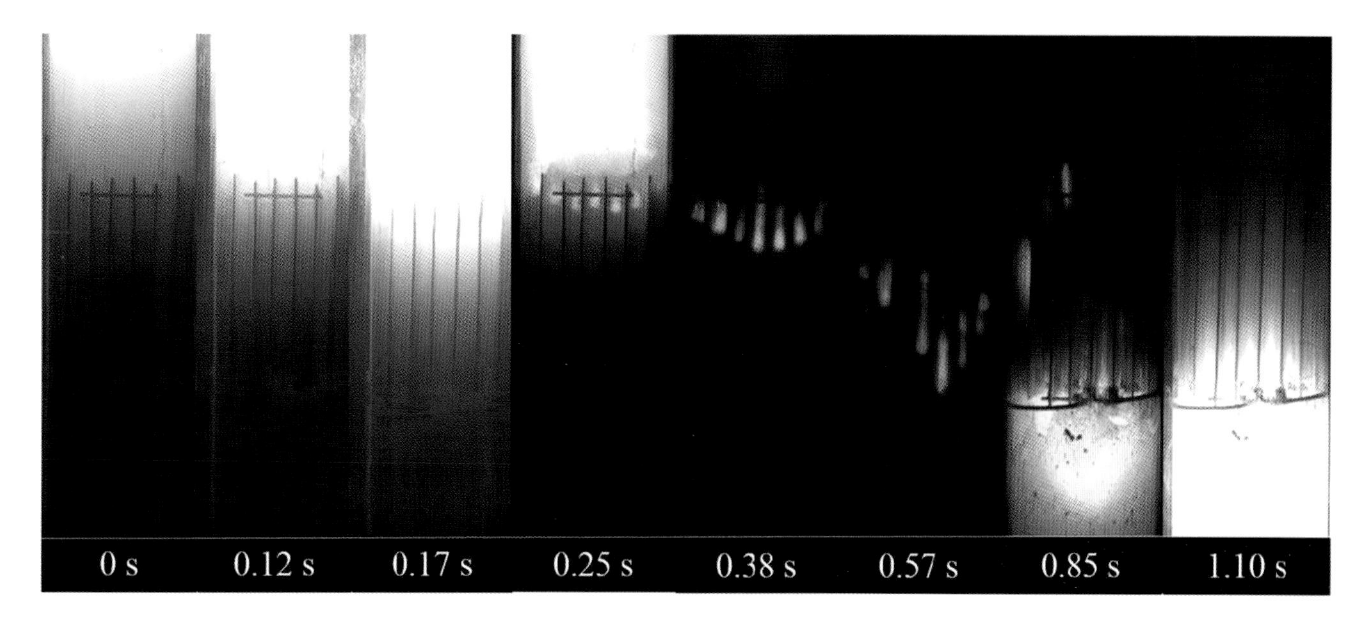

Figure 1.28

1.7.5 Aluminum–Methane/Air Dual-Front Flame Decoupling and Recoupling

Jan Palecka, Phillippe Julien, Samuel Goroshin, Andrew Higgins, David Frost, and Jeffrey Bergthorson

McGill University

Series of frames illustrating the propagation, quenching, and reestablishment of dual-front flames in a suspension of aluminum particles in a premixed methane-oxidizer mixture. The experiment is performed in tubes of diameter of 48 mm, within which are placed a set of quenching channels, formed by equally-spaced plates. The tests are performed in mixtures of aluminum suspensions (d32 = 5.6 μm) in a methane/oxygen/nitrogen gas (16.3% O2/8.1% CH4/75.6% N2). Upon reaching the small channels, the aluminum-methane flame decouples and only the dim orange front, corresponding to the methane-oxygen reaction, propagates through the channels. Soon after the flame emerges back into the tube, the aluminum front forms again in the wake of the methane front and both fronts thermally re-couple. The aluminum concentration in the image is about 400 g/m3.

Supported by Natural Sciences and Engineering Research Council of Canada and Martec, Ltd. (Collaborative Research and Development Grant) and Defense Threat Reduction Agency under contract HDTRA 1-11-1-0014.

Keywords:
Aluminum–methane/air; dual front; decoupling; recoupling

Reference

Palecka J, Julien P, Goroshin S, Bergthorson JM, Frost DL, Higgins AJ. Quenching distance of flames in hybrid methane–aluminum mixtures. Proc Combust Inst 2015;35:2463–70. http://dx.doi.org/10.1016/j.proci.2014.06.116.

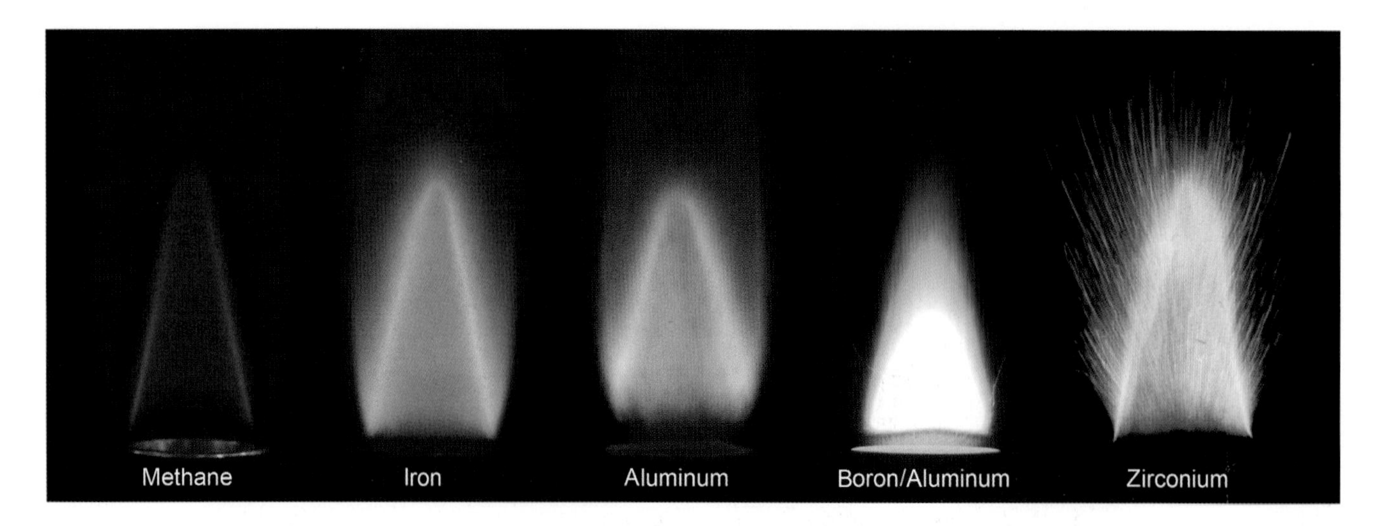

Figure 1.29

1.8 Metal Power Flames

1.8.1 Flames of Different Metal Powders Burning in Air

Michael Soo, Philippe Julien, Jan Palecka, David Frost, Samuel Goroshin, and Jeffrey Bergthorson

McGill University

The pictures show 5 different flames stabilized on a laminar metal-fuel burner. The first flame is a methane flame, similar to that stabilized on a standard Bunsen burner. Flames of various metal powders suspended in air and other oxidizing gases are shown in the following four frames.

Support for this work was provided by the Natural Sciences and Engineering Research Council of Canada.

Keywords:
Metal powders/air

Reference

JM Bergthorson, S Goroshin, MJ Soo, P Julien, J Palecka, DL Frost, and DL Jarvis. Direct combustion of recyclable metal fuels for zero-carbon heat and power. Applied Energy 160 (2015), 368–382.

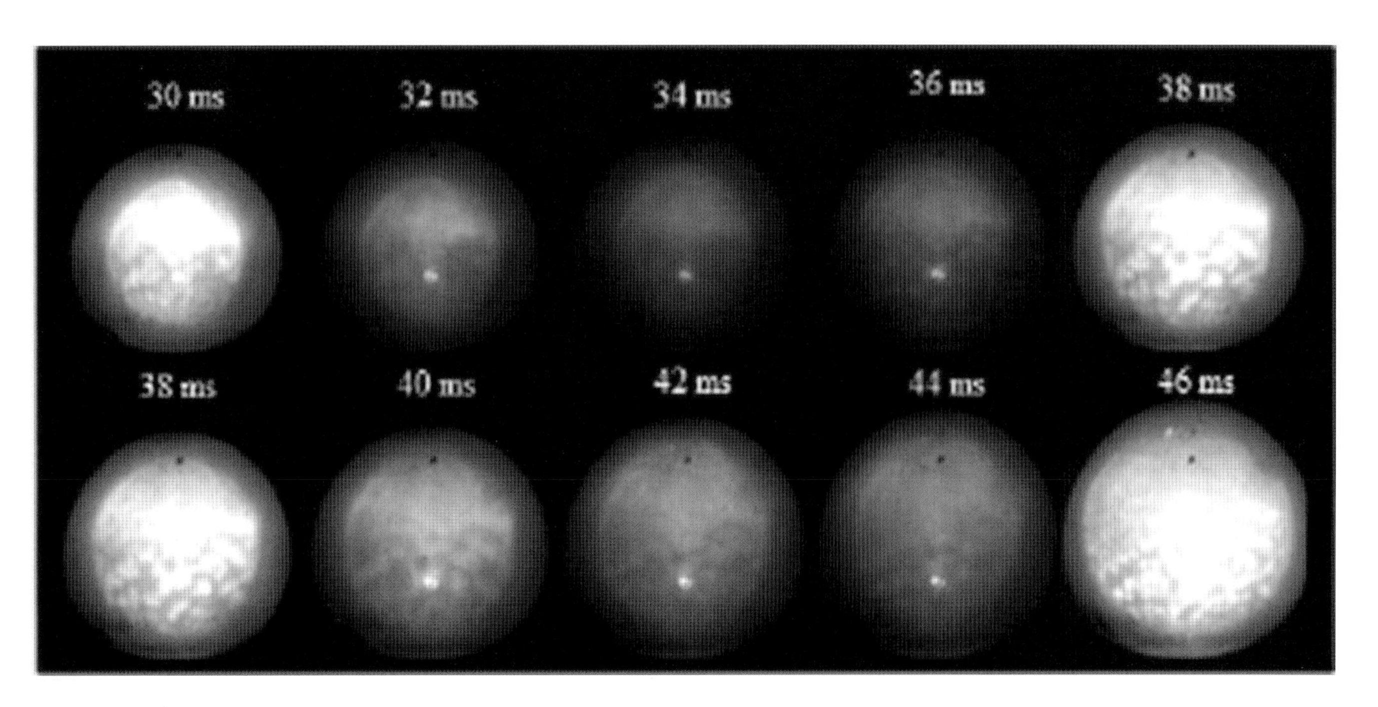

Figure 1.30

1.8.2 Pulsating Aluminum Flame

Philippe Julien, James Vickery, Samuel Goroshin, David Frost, and Jeffrey Bergthorson

McGill University

The pictures show two periods of a pulsating aluminum flame. The laminar flame is freely-propagating inside a 30 cm latex balloon. The pulsations arise from thermo-diffusive instabilities. The flame is propagating in a gaseous mixture of 40% oxygen and 60% argon. The aluminum concentration is on the order of 300 g/m3, which corresponds to an equivalence ratio of 0.5, and the particle Sauter mean diameter is 5.6 μm.

Supported by Natural Sciences and Engineering Research Council of Canada and Martec, Ltd. (Collaborative Research and Development Grant) and Defense Threat Reduction Agency under contract HDTRA 1-11-1-0014.

Keywords:
Aluminum/air, pulsating

References

Julien P., Vickery J., Whiteley S., Wright A., Bergthorson J. M., Frost D. L., "Effect of scale on freely propagating flames in aluminum dust clouds", J. Loss Prev. Process Ind. 36, 230–236.

Julien, P., Vickery, J., Goroshin, S., Frost, D. L. & Bergthorson, J. M., "Freely-propagating flames in aluminum clouds". Combust. Flame 162, 4241–4253 (2015).

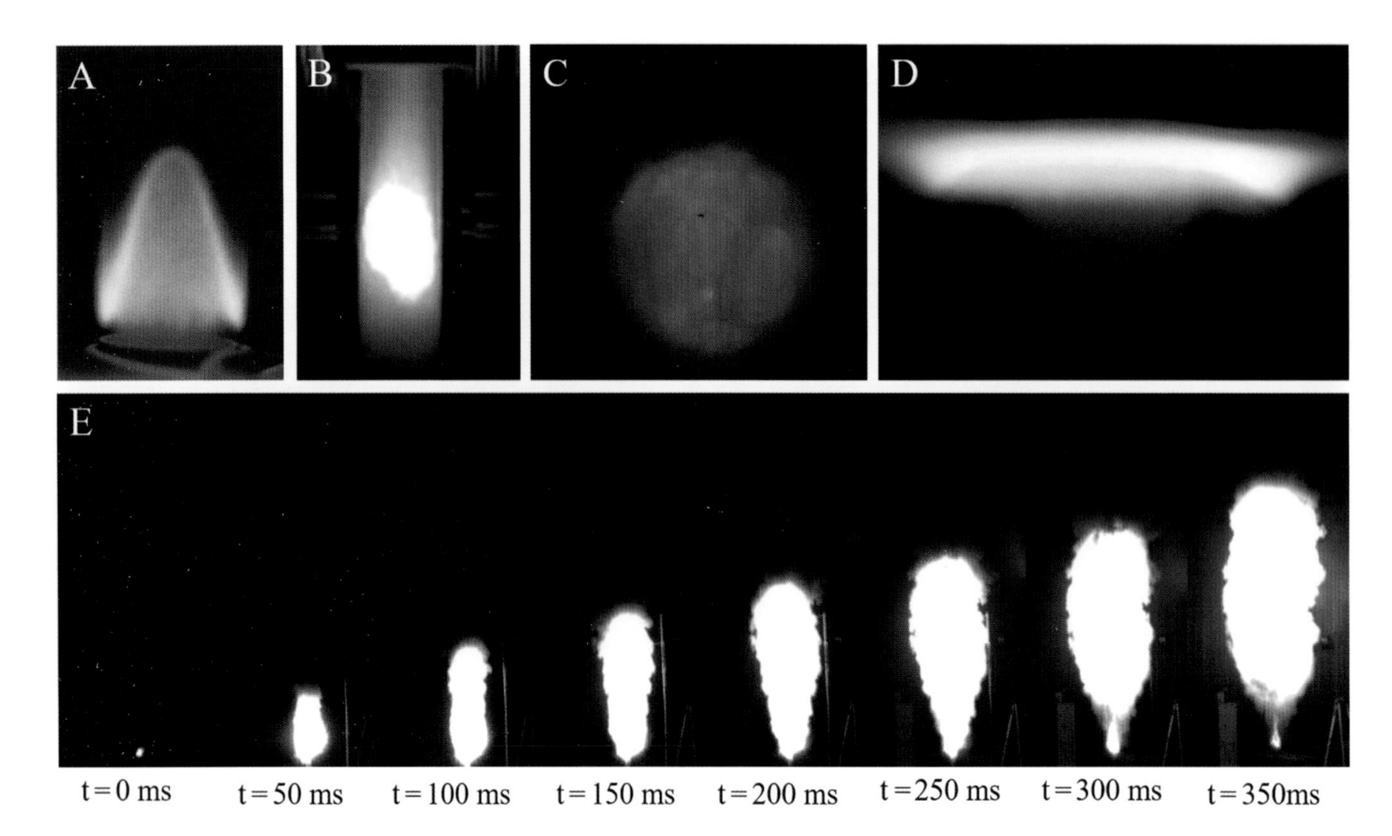

Figure 1.31

1.8.3 Aluminum/Air Flames

Philippe Julien, James Vickery, Michael Soo, David Frost, Samuel Goroshin, and Jeffrey Bergthorson

McGill University

The pictures show 5 different aluminum flames. In all cases, the particle size is around 6 μm, and all flames are fuel rich (equivalence ratio greater than one). The figures show pure aluminum-air flames, in different geometric configurations. The configurations are the following: a) Bunsen flame, b) aluminum flame in a tube, c) spherically-propagating aluminum flame, d) flat, counter-flow aluminum flame, e) Still frames of a large-scale (2 m width, 4 m height), cylindrical aluminum flame.

Support for this work was provided by the Natural Sciences and Engineering Research Council of Canada and Martec, Ltd., under a Collaborative Research and Development Grant, and the Defense Threat Reduction Agency under contract HDTRA1-11-1-0014.

Keyword:
Aluminum/air

References

Palecka, J. et al. Quenching distance of flames in hybrid methane–aluminum mixtures. Proc. Combust. Inst. 35, 2463–2470 (2015).

Julien, P., Vickery, J., Goroshin, S., Frost, D. L. & Bergthorson, J. M. Freely-propagating flames in aluminum clouds. Combust. Flame 162, 4241–4253 (2015).

Julien, P. et al. Flame speed measurements in aluminum suspensions using a counterflow burner. Proc. Combust. Inst. 36, 2291–2298 (2017).

Julien, P. et al. Effect of scale on freely propagating flames in aluminum dust clouds. J. Loss Prev. Process Ind. 36, 230–236 (2015).

2 Computational Fluid Dynamics

Edited by Mark Vaccari

Introduction

Mark Vaccari

Humans have an insatiable desire to model physical phenomena and to continuously improve those models, be it modeling weather patterns for forecasting, molecular modeling in pharmaceutical research, or even biscuits baking in an oven.[1] Of particular interest, especially given the nature of this book, is the modeling of combustion and combustion systems. Modeling of combustion comes in many forms, ranging from simple stoichiometry of global reactions, to detailed kinetic modeling of elementary reactions in a combustion mechanism, to one-dimensional reactor models, to full-blown transient three-dimensional models using Computational Fluid Dynamics (CFD).

CFD is an area of fluid mechanics that deals with the computer simulation of fluid flow, heat transfer, combustion, etc. CFD is rooted in solving the fundamental governing equations for continuity, momentum, and energy. The volume of interest is discretized into a finite number of volume elements called cells. A single model may have many millions, or even billions, of cells. Appropriate models for turbulence, combustion, radiation, etc. are selected. The knowledge of which particular models are most applicable for a given simulation is a skill acquired through both schooling and experience running CFD simulations. Boundary conditions are then applied to the various boundaries in the model. Then the governing equations in each cell are solved. This is the most computationally intensive task, as all of the equations are coupled to each other in a complicated and nonlinear fashion. Even mid-sized steady-state models, with 20–30 million cells, can take several days on a large computing cluster to solve. If, instead of running the models on a computing cluster, they were run on a single desktop, these models could take more than a year to solve! For larger models and/or transient models (time varying), the computational requirements are even larger, often requiring tens of thousands or even hundreds of thousands of core-hours to solve.

CFD simulations are increasingly becoming a mainstay in the analysis of combustion systems, both in academic/research settings as well as industrial settings. Combustion CFD simulations allow for deeper insight into systems – solving for velocity, temperature, pressure, and species concentrations everywhere in the volume instead of just at a few probe points with experimental apparatus. Additionally, it allows for full-scale simulation of equipment, where full-scale testing is typically prohibitively expensive and challenging (if possible at all). Finally, because combustion CFD simulations predict variables in all locations, it allows for the prediction of quantities that cannot be measured in an experimental apparatus (e.g., vorticity).

The following images demonstrate a variety of applications of combustion CFD simulations from a diverse set of contributors. The images correspond loosely to the following categories (in order): gas turbines, burners in isolation, rocket engines, industrial applications of combustion, emissions reduction, and explosions.

Reference

1. D. Fahloul, G. Trystram, A. Duquenoy, I. Barbotteau, Modelling heat and mass transfer in band oven biscuit baking, *LWT – Food Science and Technology*, 27, 2 (1994), 119–124, ISSN 0023-6438, https://doi.org/10.1006/fstl.1994.1027.

Figure 2.1

2.1 LES Simulation of a Gas Turbine Combustor

Niveditha Krishnamoorthy, Chandraprakash Tourani, and Matthew Godo
Siemens Product Lifecycle Management Software Inc.

Artyom Pogodin
OOO Siemens Industry Software

Karin Fröjd
Siemens Industry Software AB

Transient LES simulation of turbulent combustion in a steadily burning annular gas turbine combustor, using Simcenter STAR-CCM+ from Siemens. The combustion chamber is 16.5 cm long with a height of 6 cm. The simulation is run for one sector and visually duplicated through a periodic transform. The sector is discretized using a 316,000-cell polyhedral mesh. An inlet mass flow of air at 500 K is provided through the plenum inlet, whereof a portion is reaching the combustion chamber through swirler and dilution holes, at velocities of around 100 m/s. The swirler walls and dilution holes are visualized in the illustration. JP10 is injected at 300 K at a mass flow rate of 1 g/s in the center of the swirler using a lagrangian injector. The LES WALE turbulence model is used and the mass averaged Reynolds number is 5×10^4. The Flamelet Generated Manifold combustion model is applied with FGM Kinetic Rate for flame propagation. The blue field visualizes areas with high fuel concentration, where the fuel–air mixture fraction is higher than 0.18. The flame front is visualized through volume rendering, highlighting areas where the combustion progress variable lies between 0.93 and 0.96. The redder the color, the higher the progress variable.

Keywords:
LES; FGM; JP10; gas turbine; combustion; spray combustion

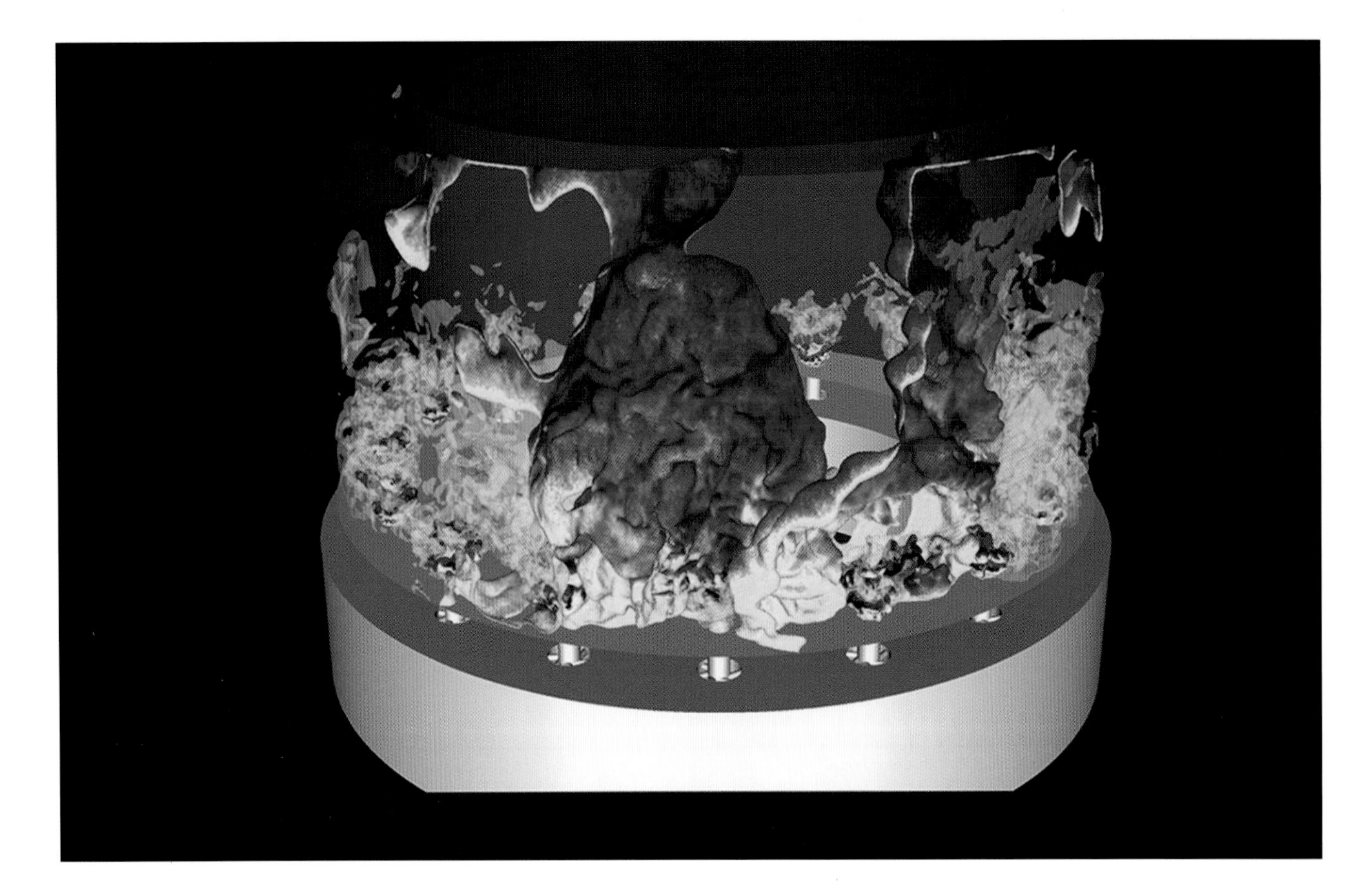

Figure 2.2

2.2 Annular Burner Ignition

Eleonore Riber and Bénédicte Cuenot
CERFACS

Keywords:
Ignition; aeronautical propulsion

This image shows an instantaneous view of the propagating flame in the annular burner experimentally studied in Ref 1. A 0.74 equivalence ratio mixture of propane and air is injected at ambient temperature and pressure through 16 swirled injectors (swirl number 0.82). Results are analyzed in detail in Ref 2.

The authors would like to acknowledge PRACE for awarding access to computing resource of TGCC (France).

References

1. J.-F. Bourgouin, D. Durox, T. Schuller, J. Beaunier, S. Candel, Combust, *Flame* 160 (2013), 1398–1413.
2. M. Philip, M. Boileau, R. Vicquelin, E. Riber, T. Schmitt, B. Cuenot, D. Durox, S. Candel, Large eddy simulations of the ignition sequence of an annular multiple-injector combustor, *Proceedings of the Combustion Institute* 35, 3(2015), 3159–3166.

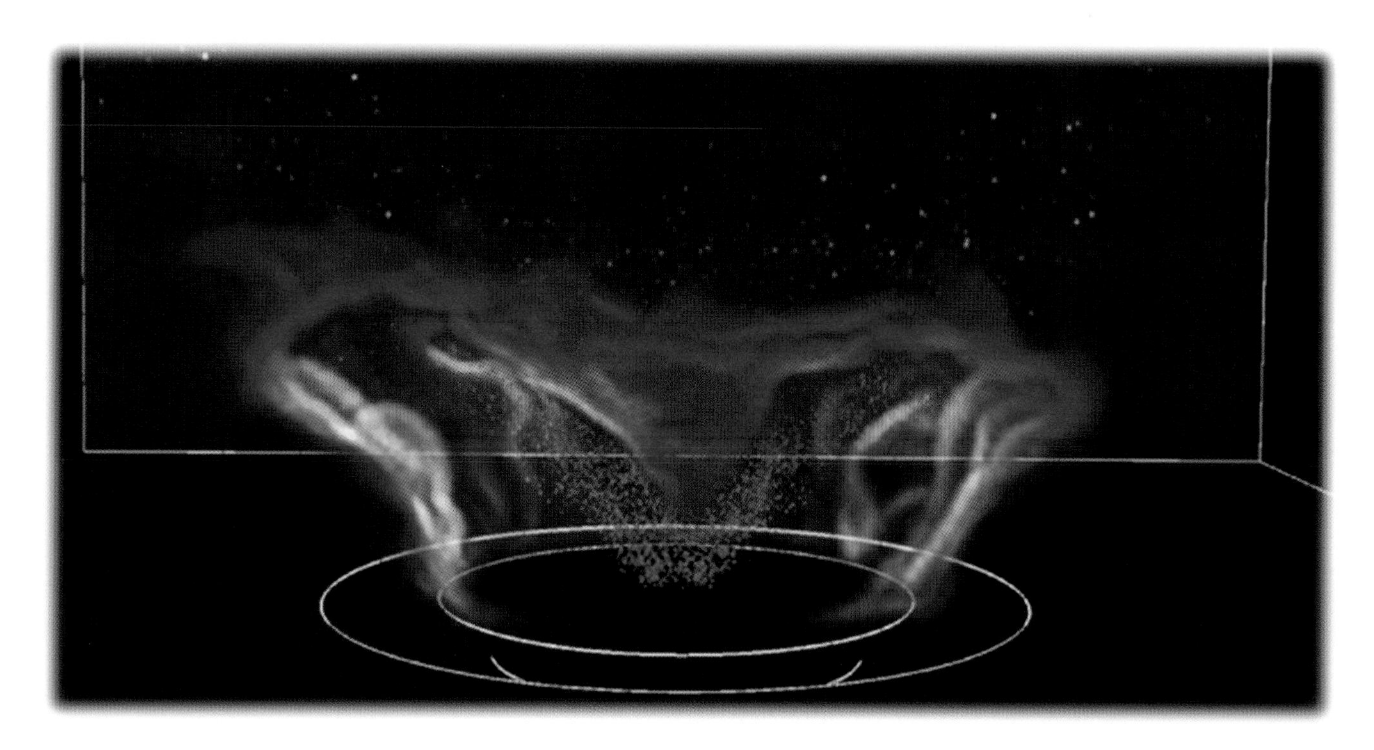

Figure 2.3

2.3 Numerical Simulation of a Spray

Damien Paulhiac, Bénédicte Cuenot, and Eleonore Riber
CERFACS

The atmospheric swirled spray flame corresponds to the experimental setup of Ref 1. The image is a visualization of the fuel droplets and the flame surface.

The liquid fuel is n-heptane, injected at a mass flow rate of 0.12 g/s and at ambient pressure and temperature. The global equivalence ratio of the burner is 0.17.

The resulting flame has a complex, partially premixed structure. In addition, a non-negligible part of droplets is able to cross the flame front and ignite in the burnt gas to form individually burning droplets.

The support of SAFRAN is acknowledged.

Keywords:
Spray flames; aeronautical propulsion

Reference

1. D. E. Cavaliere, J. Kariuki, E. Mastorakos, A comparison of the blow-off behaviour of swirl-stabilized premixed, non-premixed and spray flames, *Flow, Turbulence and Combustion*, 91 (2013), 347–372.

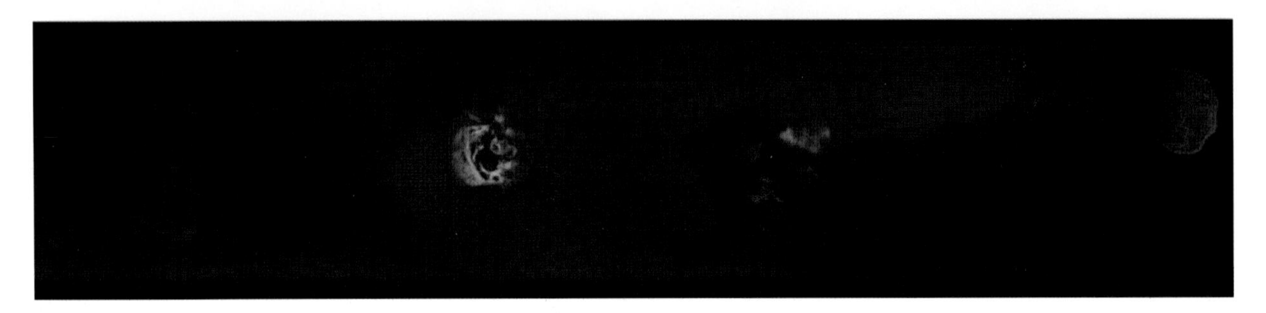

Figure 2.4

2.4 Flame Ignition and Propagation in Aeronautical Swirled Multi-burners

David Barre, Lucas Esclapez, Eleonore Riber, and Bénédicte Cuenot

CERFACS

The experimental configuration was designed by CORIA in the context of the European project KIAI (Knowledge for Ignition, Acoustics and Instabilities – 7th Framework Program – 2009/2013). The setup allows to vary the separation distance between the burners. Methane and air mass flow rates are, respectively, 0.192 g/s and 5 g/s for each individual injector, leading to a global equivalence ratio of 0.66. The radial swirler is composed of 18 vanes inclined at 45°, leading to a swirl number of 0.76. The burner operates at ambient conditions. The images show a visualization of the flame surface from a side view (top image) and top view (bottom image). Results show the importance of burner spacing on the ignition sequence and the partially premixed flame structure.

The research leading to these results has received funding from the European Community's 7th Framework Program (FP7/2007–2013) under Grant Agreement No. ACP8-GA-2009-234009. This work was granted access to the HPC resources of [CCRT/IDRIS] under the allocation 2013-x20132b5031 made by GENCI (Grand Equipement National de Calcul Intensif). This research is also part of a 2013 INCITE award of the Department of Energy, and used resources of the Argonne Leadership Computing Facility at Argonne National Laboratory, which is supported by the Office of Science of the U.S. Department of Energy under contract DE-AC02–06CH11357.

Keywords:

Large eddy simulation; ignition; turbulent partially premixed flame; flame propagation; multi-injectors

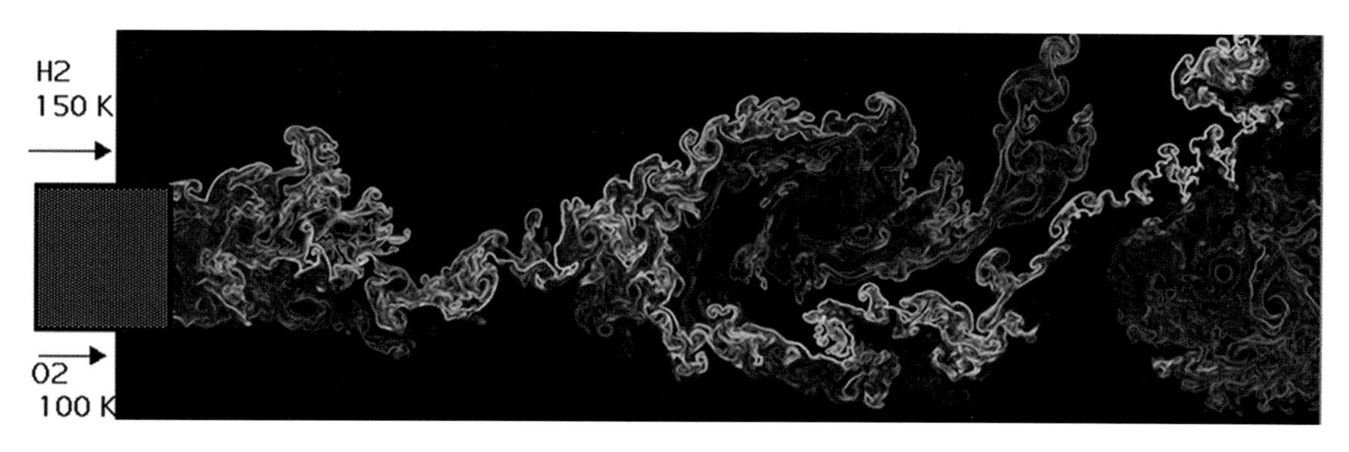

Figure 2.5

2.5 Direct Numerical Simulation of a Transcritical Flame

Anthony Ruiz, Raphaël Mari, and Bénédicte Cuenot

CERFACS

Laurent Selle

IMFT

Thierry Poinsot

IMFT/CERFACS

The image, colored by scalar dissipation rate, shows a cryogenic diffusion flame of H_2/O_2 attached to a lip of thickness 0.5 mm. Hydrogen is injected at 150 K while oxygen is injected at 100 K in a supercritical state. The density ratio between the two reactant streams is 80. The chamber pressure is 100 bar. Results are detailed in Ref 1.

Support for this research was provided by Snecma and CNES. The simulations used resources of the high-performance computing resources of CINES under the allocation 2011-c2011025082 made by GENCI.

Keywords:

Cryogenic liquid rocket engine; supercritical flames; direct numerical simulation

Reference

1. A. M. Ruiz, G. Lacaze, J. C. Oefelein, R. Mari, B. Cuenot, L. Selle, T. Poinsot, Numerical benchmark for high-Reynolds-number supercritical flows with large density gradients, *AIAA Journal* 54, 5(2015), 1445–1460.

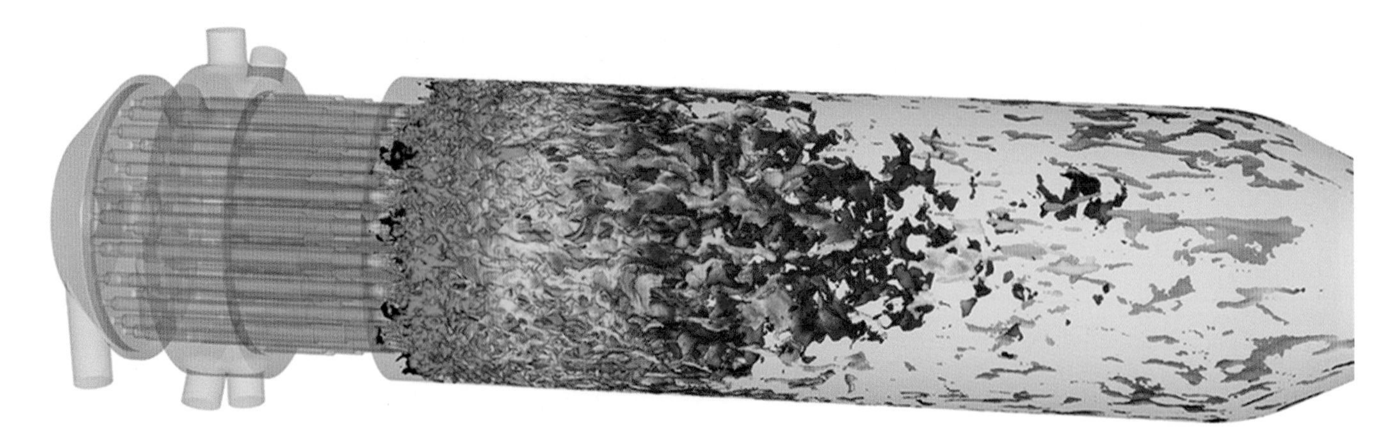

Figure 2.6

2.6 Large Eddy Simulation of a 42-Injector Liquid Rocket Engine

Annafederica Urbano and Laurent Selle
IMF Toulouse

Gabriel Staffelbach and Bénédicte Cuenot
CERFACS

Thomas Schmitt and Sebastien Ducruix
Centrale-Supelec

The image shown here is a temperature isosurface colored by velocity of the BKD operated at the P8 test facility at DLR Lampoldshausen (Ref 1). It is an H_2/LOx burner, operating at 70 bar with cryogenic propellants injected in transcritical (LOx) or supercritical (H_2) states. The oxidizer/fuel ratio is 4. In such configuration, the turbulent flame is in a purely non-premixed combustion regime. The work was published in *Combustion and Flame* (Ref 2).

This investigation was carried out in the framework of the French-German REST program initiated by CNES and DLR. All geometrical, operational, and measurement data related to the BKD were kindly provided by DLR Lampoldshausen. The authors are particularly grateful to Stefan Gröning and colleagues who performed the experiments and formulated the test case. Support provided by Safran (Snecma), the prime contractor of the Ariane rocket propulsion system, is gratefully acknowledged.

The authors acknowledge PRACE for awarding them access to resource FERMI based in Italy at Cineca. This work was granted access to the high-performance computing resources of IDRIS under the allocation x20152b7036 made by Grand Equipement National de Calcul Intensif. The support of Calmip for access to the computational resources of EOS under allocation P1528 is acknowledged. The research leading to these results has received funding from the European Research Council under the European Union's Seventh Framework Programme (FP/2007-2013)/ ERC Grant Agreement ERC-AdG 319067-INTECOCIS.

Keywords:
Combustion instabilities; liquid rocket engines; large eddy simulation

References

1. S. Gröning, J. S. Hardi, D. Suslov, M. Oschwald, Injector-driven combustion instabilities in a hydrogen/oxygen rocket combustor, *Journal of Propulsion and Power* 32.3 (2016), https://doi.org/10.2514/1.B35768.
2. Annafederica Urbano, Laurent Selle, Gabriel Staffelbach, and Bénédicte Cuenot, Exploration of combustion instability triggering using large eddy simulation of a multiple injector liquid rocket engine, *Combustion and Flame* 169 (2016), 129–140.

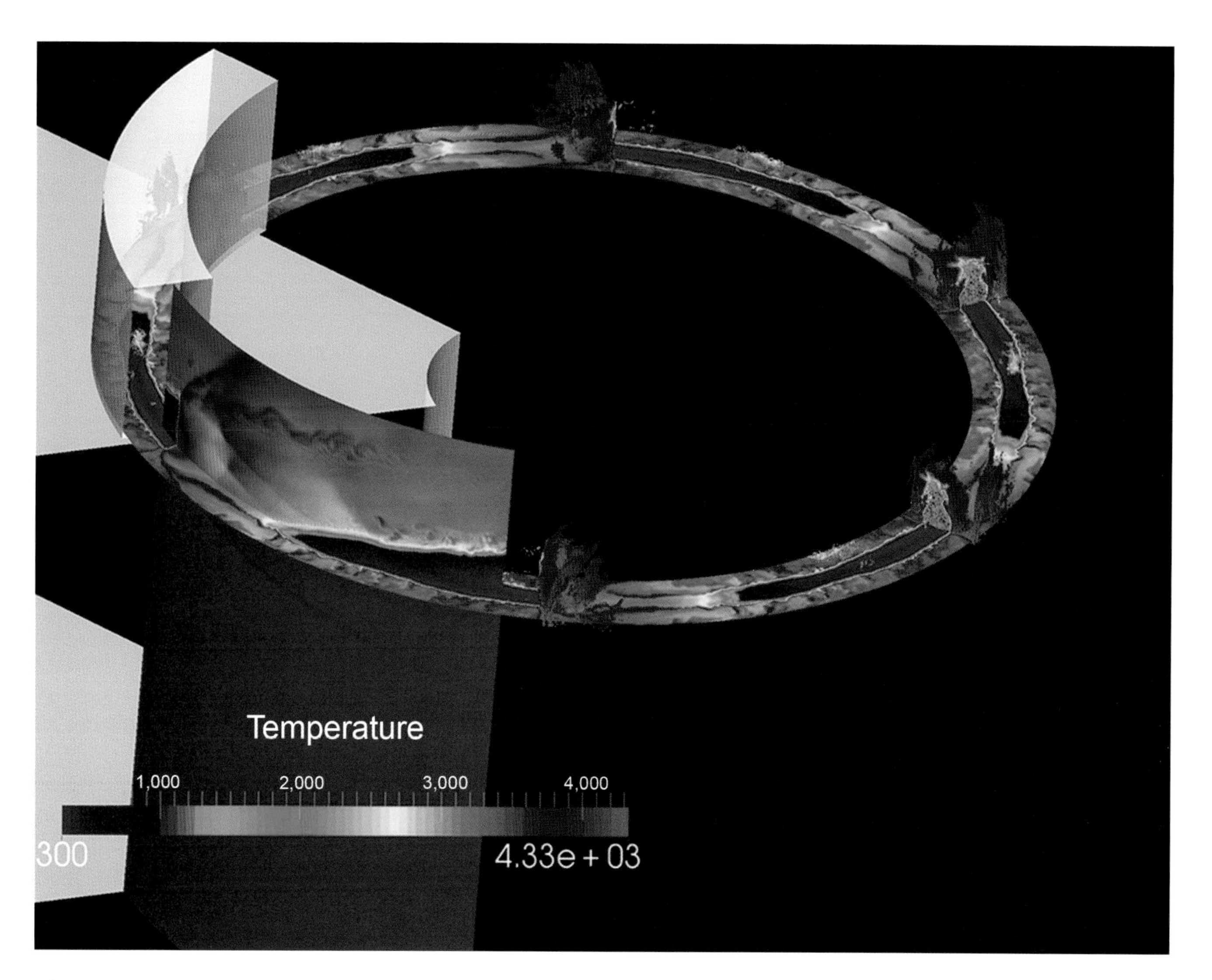

Figure 2.7

2.7 Numerical Simulation of a Rotative Detonation Engine

Romain Bizzari and Antoine Dauptain
CERFACS

Keywords:
Combustion instabilities; liquid rocket engines; large eddy simulation

The geometry is a 105 mm diameter, 20 mm high, and 10 mm thick cylinder, as in Ref 1. A stoichiometric H_2/O_2 mixture is injected at the bottom at P_{total} = 10 bars and T_{total} = 300 K, and feeds six propagative detonation fronts. The exhaust is maintained at 0.4 bar. The isosurface of pressure colored by temperature shows the six detonative fronts. Shocks and mixing layer can be observed in the hot gases.

Reference

1. Yohann Eude, Dmitry Davidenko, Francois Falempin, and Iskender Gökalp, Use of the adaptive mesh refinement for 3D simulations of a CDWRE (continuous detonation wave rocket engine). AIAA paper 2236 (2011), 2011.

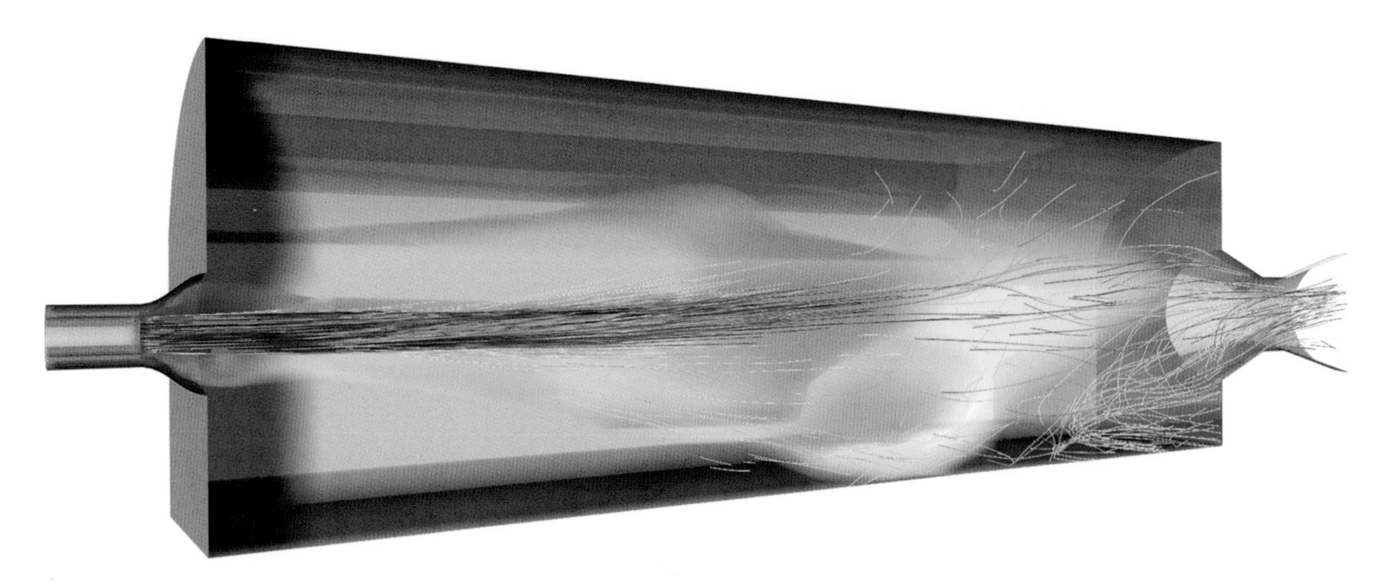

Figure 2.8

2.8 RANS Simulation of an IFRF Coal Furnace

Steve Evans

Siemens Industry Software Computational Dynamics Limited

Piyush Thakre

Siemens Product Lifecycle Management Software Inc.

Karin Fröjd

Siemens Industry Software AB

Steady-state RANS simulation of IFRF Coal Furnace (Ref 1) using Simcenter STAR-CCM+ from Siemens. The furnace is 6.5 m long, 2 m high, and 2 m wide. Pulverized coal of temperature 303 K is delivered at a mass flow rate of 0.104 kg/s and a velocity of 50 m/s with a transport air flow of 343 K at 75 m/s through a coal gun of inner diameter 11.5 cm and outer diameter 14 cm. An outer swirling co-flow of air at 300K is delivered through an annulus of inner diameter of 14 cm and outer diameter 23.3 cm at an axial velocity of 30–50 m/s and a swirl number of 1.16. The flow is turbulent with a mass averaged Reynolds number of 10^4. The k–ε turbulence model is used with lagrangian injection of coal particles, using a Rosin–Rammler particle-size distribution with a reference diameter of 0.2 mm. The particles react through devolatilization and char oxidation by O_2, H_2O, and CO_2. The secondary gas phase reactions are modeled through the Eddy Break-Up combustion model. Radiation is accounted for through participating media radiation (DOM) with gray spectrum model. The red-yellow field indicates volume of significant CO mass fraction ($>$ 1% by mass). Redder color indicates higher CO concentration. It can be noted that significant CO concentrations exist at the end of the furnace. The blue lines show tracks of large coal particles (particle diameter $>$ 0.15 mm). Darker blue indicates larger particles. As can be seen, most particles are burnt out in the main flame and in the recirculation, but some particles escape as large particles, which is easy to see through this visualization. Those are the particles that travel more or less straight between the injector and the outlet and thus do not fully burn out, as well as particles with very high initial diameter, by outlet consisting of primarily ash (initial ash content is 8%).

Keywords:
Coal; combustion; furnace; CFD; RANS; IFRF

Reference

1. André A. F. Peters and Roman Weber, Mathematical modeling of a 2.4 MW swirling pulverized coal flame, *Combustion Science and Technology* 122 (1997), 131–182.

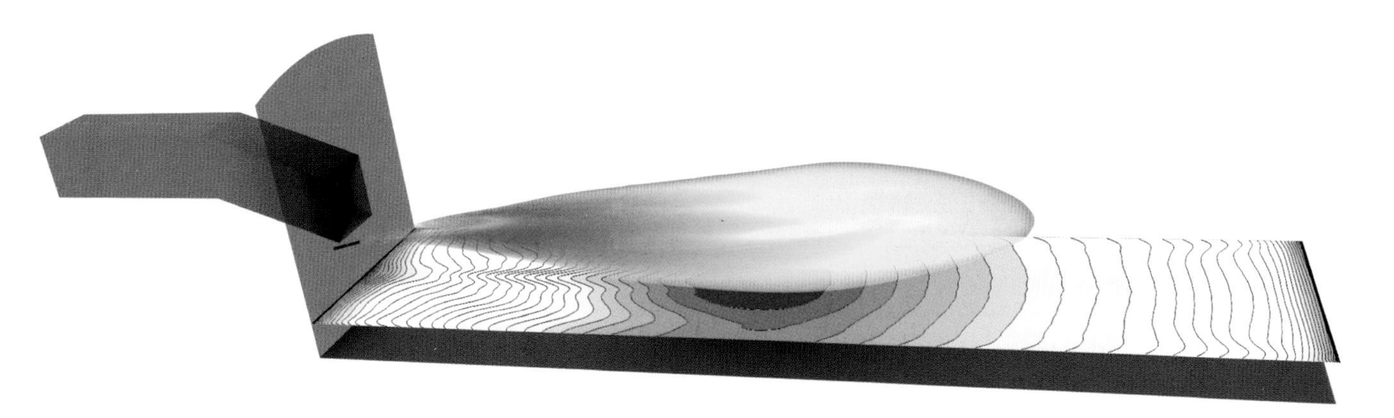

Figure 2.9

2.9 RANS Simulation of a Glass Furnace

George Mallouppas
Siemens Industry Software Computational Dynamics Limited

Rajesh Rawat
Siemens Product Lifecycle Management Software Inc.

Karin Fröjd
Siemens Industry Software AB

Steady-state RANS simulation of IFRF glass furnace (Ref 1) with a molten glass layer added at the base, using Simcenter STAR-CCM+ from Siemens. The furnace is 3.8 m long, 1.105 m high, and 0.88 m wide. Air is delivered at a velocity of 125 m/s and a temperature of 1,373 K through a 27.8 cm × 27.2 cm rectangular air inlet. Methane is delivered at a velocity of 10 m/s and a temperature of 283 K through a circular nozzle of 1.2 cm diameter, giving a global mixture with 10% excess air. The geometry is discretized using a 1,000,000-cell hexahedral mesh with local refinement by the fuel injection and the flame. Only half of the furnace is simulated using symmetry boundary conditions along the centerplane, and the picture is created using a symmetry transform. The flow is turbulent with a mass averaged Reynolds number of 8×10^5. The SST k–ω turbulence model is used with the complex chemistry combustion model, applying a skeletal 17 species mechanism with 73 reactions from Lu et al. (Ref 2). Radiation is accounted for through participating media radiation (DOM) with gray spectrum model. A molten glass layer of 15 cm thickness is included at the base of the combustion chamber using the VOF model. The red-yellow field visualizes the flame by showing volumes where temperature is higher than 2,000 K. The higher the temperature, the redder the color. The blue-red field indicates incident radiation at the molten glass surface, ranging from 2 MW/m^2 in the dark blue areas to 3.4 MW/m^2 in the red area just below the flame. It is easy to see where the incident radiation is lowest and consequently where there is highest risk for the glass to solidify.

The very same case but without the molten glass layer is presented in detail in Mallouppas et al. (Ref 3).

Keywords:
Glass furnace; combustion; methane; RANS; CFD

References

1. T. Nakamura, W. L. Vandecamp, J. P. Smart, Further studies on high temperature gas combustion in glass furnaces, Technical report, IFRF Doc No F 90/Y/7, 1990.
2. R. Sankaran, E. R. Hawkes, J. H. Chen, T. F. Lu, C. K. Law, Structure of a spatially developing turbulent lean methane–air Bunsen flame, *Proceedings of the Combustion Institute* 31 (2007), 1291–1298.
3. G. Mallouppas, Y. Zhang, R. Rawat, Modelling of combustion, NO_x Emissions and radiation of a natural gas fired glass furnace, AFRC 2014 Industrial Combustion Symposium.

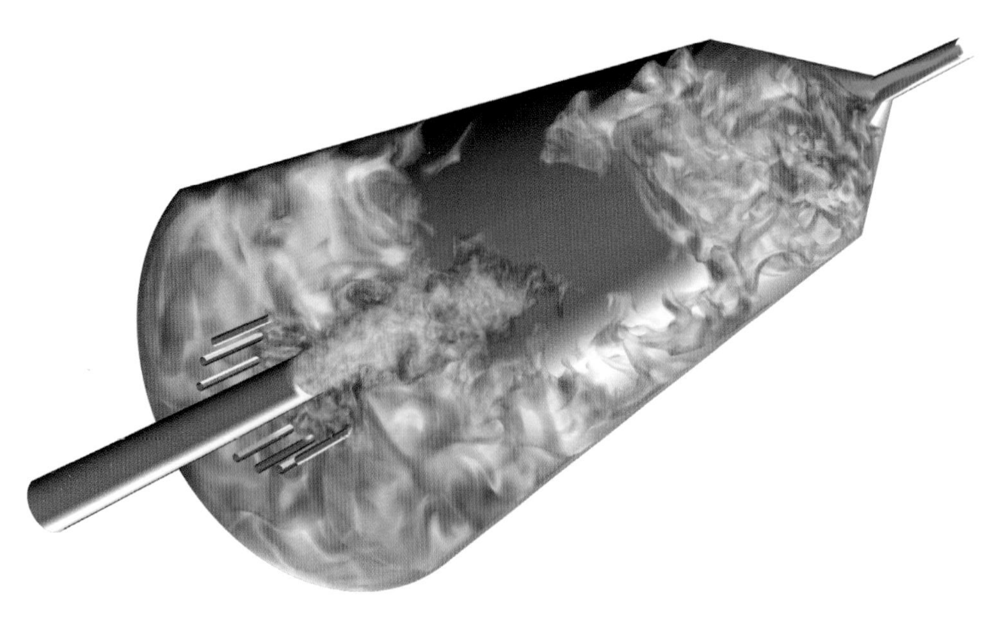

Figure 2.10

2.10 LES Simulation of a Flameless Combustor

Carlo Locci

Siemens Industry Software GmbH

George Mallouppas

Siemens Industry Software Computational Dynamics Limited

Rajesh Rawat

Siemens Product Lifecycle Management Software Inc.

Karin Fröjd

Siemens Industry Software AB

Transient LES simulation (Ref 1) of a flameless combustor from Verissimio et al. (Ref 2) in Simcenter STAR-CCM+ from Siemens. Methane is burnt in a cylindrical combustion chamber of diameter 10 cm and length 34 cm, with a convergent nozzle of length 15 cm by the end. Methane is injected at 0.2 g/s at 293 K through 16 fuel inlets of diameter 2 mm surrounding a central air inlet of diameter 10 mm. The air is preheated to 673 K and injected at a mass flow of 4.6 g/s. The flow is highly turbulent with a mass averaged Reynolds number of 6×10^4. LES with WALE sub-grid scale model is used to model turbulence. The combustion model is Flamelet Generated Manifold with kinetic rate for flame propagation. Thermal and prompt NO_x are calculated. The blue field highlights fuel inlets by visualizing mixture fractions > 0.05. The orange field highlights the main combustion zone by showing areas where the combustion progress variable lies between 0.05 and 0.95. The redder the color, the higher the progress variable. The green areas highlight areas where dry NO_x is > 50 ppm by mass. Here we can see that NO_x is formed in the recirculation regions, where the residence time is high enough to allow for significant NO_x production. We can also see that the residence time in the reactor overall is high enough to generate $NO_x > 50$ ppm, indicated by the green volumes by the exhaust of the reactor.

Keywords:
Flameless combustor; combustion; LES; CFD; NO_x

References

1. C. Locci, G. Mallouppas, R. Rawat, Numerical analysis of NO and CO in a flameless burner, AFRC 2016 Industrial Combustion Symposium.
2. A. S. Verissimo, A. M. A. Rocha, M. Costa, Operational, combustion, and emission characteristics of a small-scale combustor, *Energy & Fuel* 25 (2011), 2469–2480.

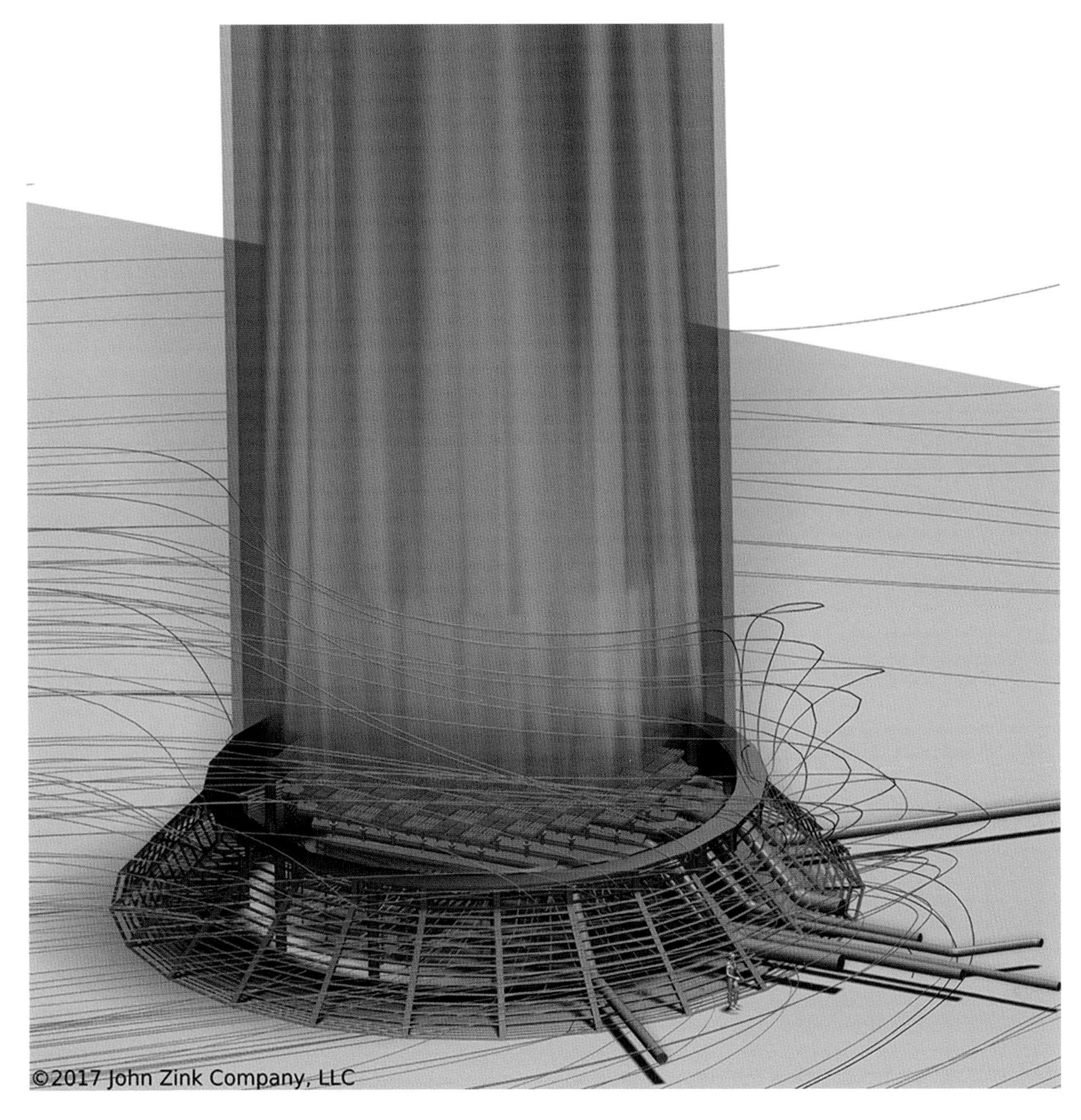

Figure 2.11

2.11 Enclosed Ground Flare

Mark Vaccari
John Zink Hamworthy Combustion

Enclosed ground flares are used to supplement, and sometimes replace, elevated flares in the safe disposal of waste gases. Enclosing the flame has several advantages: no thermal radiation outside the enclosure, low noise, and no visible flame.

This image shows results from a computational fluid dynamics simulation of an enclosed ground flare burning ethene. The flame zone is colored by temperature using volume rendering. Air streamlines were added to help visualize how air entered the unit. A model of a person is added to visualize the scale of the unit.

Keywords:
Enclosed ground flare; industrial combustion; flare

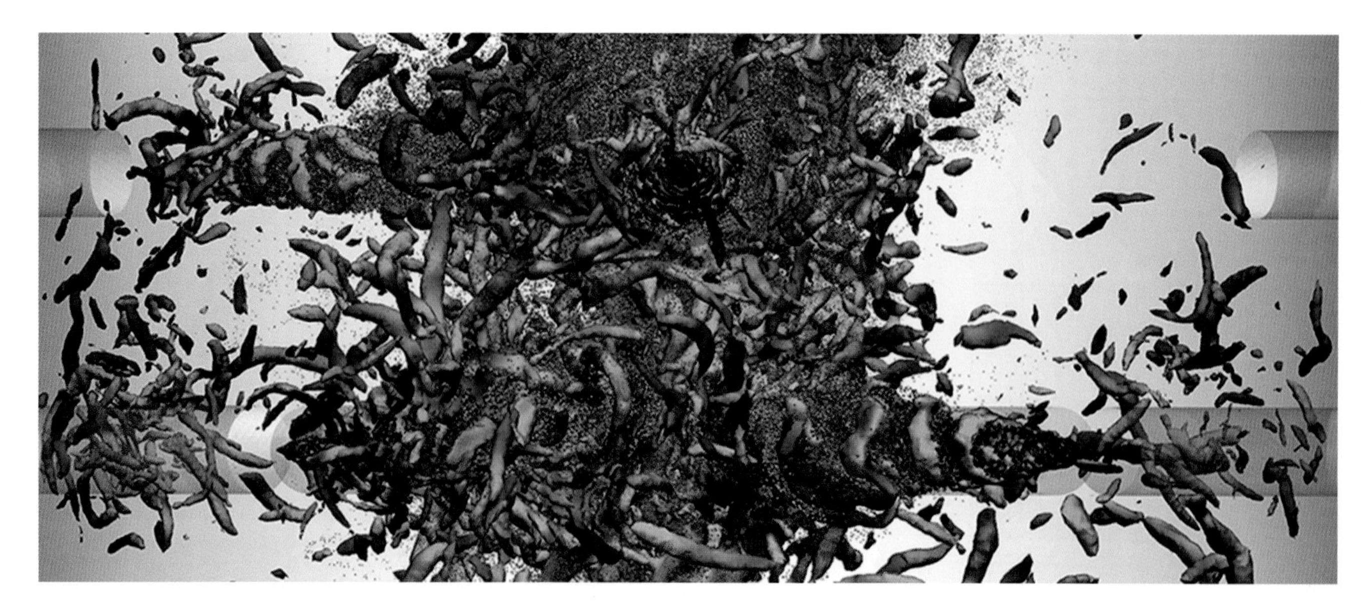

Figure 2.12

2.12 Control of Nitric Oxides Emissions

Benjamin Farcy and Luc Vervisch
INSA Rouen Normandie

Pascale Domingo
CORIA CNRS

A spray of ammonia discharges into burnt gases downstream of non-premixed burners in a 15 MW furnace, to transform nitric oxides molecules into neutral products. Multi-physics large eddy simulation is performed with a spatial resolution of 170 μm. The figure is a close-up view of the spray injection; complex vortex rings are visible on the side of the spray jets, showing the dynamics of the vortices controlled by the shear layers surrounding the liquid spray. The spray injection is conical, with an angle of 30 degrees and a dispersion angle of 4 degrees. A total number of droplets of about 600 million evolve in the simulation. The droplets are colored by their velocity magnitude (0–50 m/s from blue to red), and the visualization of the vortices is colored by the fluctuations of temperature (0–300 K from blue to red).

This work is funded by ANRT (Agence Nationale de la Recherche et de la Technology) and SOLVAY under the CIFRE no. 73683; part of it was conducted during the 2014 Summer Program of the Center for Turbulence Research, Stanford. This work was granted access to the HPC resources of IDRIS under the allocation 2014-020152 made by GENCI (Grand Equipement National de Calcul Intensif).

Keywords:
Nitric oxides emissions; direct numerical simulation; large eddy simulation

Reference

1. B. Farcy, L. Vervisch, P. Domingo, Large eddy simulation of selective non-catalytic reduction (SNCR): a downsizing procedure for simulating nitric-oxide reduction units, *Chemical Engineering and Science* 139 (2016), 285–303.

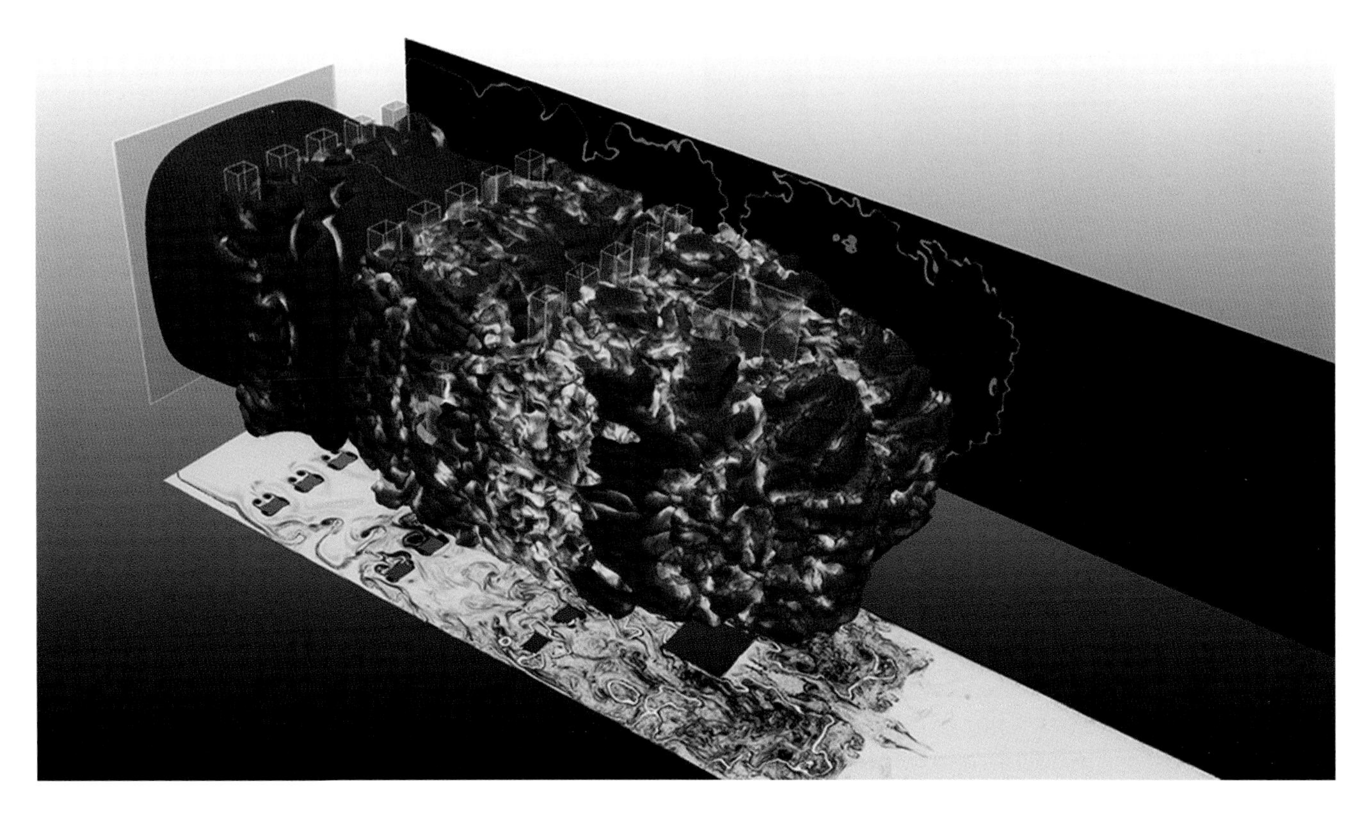

Figure 2.13 Computer model of a gas explosion in a vented chamber (the red indicates isotherms and the blue indicates vorticity).

2.13 Gas Explosion in a Medium-Scale Vented Chamber

David Barre and Olivier Vermorel

CERFACS

The medium-scale, 1.5 m-long chamber is initially filled with a stoichiometric methane–air mixture. After ignition, a laminar premixed flame develops first, which accelerates and transitions to a turbulent flame after crossing the obstacles.

The support of TOTAL is greatly acknowledged. Computational resources were provided through a grant awarded by the INCITE program of the US Department of Energy.

Keywords:
Gas explosion; large eddy simulation; turbulent combustion

Figure 2.14

2.14 Large-Scale Structure–Fire Interaction: National Fire Research Laboratory Commissioning Test, Experiment, and Modeling

Randall J. McDermott, Glenn Forney, Matthew S. Hoehler, Matthew Bundy, Lisa Choe, and Chao Zhang

NIST

Christopher M. Smith

Berkshire Hathaway Specialty Insurance

The photograph on the left shows a large-scale experiment studying the interaction between fire and mechanically loaded building elements performed during the commissioning of the National Fire Research Laboratory (NFRL) at the National Institute of Standards and Technology (NIST) in Gaithersburg, Maryland, USA (Ref 1). The midspan section of a 6.2 m-long W16×26 structural steel beam is exposed to a 700 kW open flame from the 1 m^2 natural gas burner located 1.1 m below the bottom flange of the beam. The purple-blue hue in the photo is caused by high-intensity, near-ultraviolet lighting used to illuminate the beam through the flames. The pattern of dots on the beam is imaged using two scientific cameras, and the images are processed using Digital Image Correlation (DIC) to resolve rigid body motion and deformation of the beam.

The image on the right shows a large-eddy simulation by the NIST Fire Dynamics Simulator (FDS) (Ref 2) of the NFRL commissioning test shown in the left image. The FDS results are visualized using Smokeview (Ref 3), a data visualization companion to FDS, also developed at NIST. The resolution of the simulation is 1 cm. The flame is depicted as a volume rendering of local heat release rate above a cutoff of 200 kW/m^3. False color contours of adiabatic surface temperature are shown on the I-beam. This boundary condition is taken as input to a finite-element structural analysis code to predict the deformation of the I-beam under load in a realistic fire scenario (Ref 4). The FDS code is an open-source, explicit, low-Mach flow solver. Details of the solver may be found in Ref 2.

This work was funded by NIST STRS resources under the Fire Risk Reduction in Buildings Program.

Keywords:

Structure–fire interaction; large eddy simulation; fire dynamics simulator; National Fire Research Laboratory

References

1. L. Choe, S. Ramesh, M. Hoehler, M. Bundy, M. Seif, C. Zhang, J. Gross, National Fire Research Laboratory commissioning project: testing steel beams under localized fire exposure, NIST Technical Note TN 1977, National Institute of Standards and Technology, Gaithersburg, MD, 2017.
2. K. McGrattan, S. Hostikka, R. McDermott, J. Floyd, C. Weinschenk, K. Overholt, *Fire Dynamics Simulator, Technical Reference Guide, Volume 1: Mathematical Model.* National Institute of Standards and Technology, Gaithersburg, Maryland, USA, and VTT Technical Research Centre of Finland, Espoo, Finland, NIST Special Publication 1018-1, Sixth Edition, 2013.
3. G. P. Forney, Smokeview, a tool for visualizing fire dynamics simulation data, volume ii: technical reference guide. National Institute of Standards and Technology, NIST Special Publication 1017-2, Gaithersburg, Maryland, Sixth Edition, 2013.
4. C. Zhang, L. Choe, J. Gross, S. Ramesh, M. Bundy, Engineering approach for designing a thermal test of real-scale steel beam exposed to localized fire, *Fire Technology* 53, 4 (2017), 1535–1554.

3 Internal Combustion Engines and Gas Turbines

Edited by Timothy J. Jacobs

Introduction

Timothy J. Jacobs

There are only so many technologies and devices that have the same type of impact as that of the internal combustion (IC) engine. Its ubiquitous nature pervades our everyday life, many times without us even realizing it. Whether it be the spark-ignited engine driving our vehicle, the compression-ignition engine hauling food to our local grocery store, the jet engine we hear flying 38,000 feet overhead, or the gas turbine powering the laptop screen from which we read this article, internal combustion engines are quite literally intricately and irreplaceably woven into our daily lives. The internal combustion has taken on many different forms throughout its long, greater than 150-year history, but combustion has always been one of its few constants. Indeed, combustion is even in its name and helps differentiate it from other thermodynamic work devices such heat engines and fuel cells.

In spite of combustion's constancy in the internal combustion engine, the nature of this combustion has always been of interest, of fervent research, of subject of study, and of key importance to the internal combustion engine's ability to deliver power efficiently and cleanly. Indeed, to this day, there is much to learn about combustion within the internal combustion engine and how this combustion may be exploited better to achieve goals for healthy benefit to society.

A key advancement to this exercise of improving our knowledge and use of combustion has been experimental visualization of flames, species concentrations, fluid motions, and other gradients that occur during combustion within an internal combustion engine. Tremendous progress has been made to precisely measure features of in-cylinder or in-burner reactions that not only aid our understanding of the phenomena taking place but also equip confidence in the use of simulation-based tools that further lend insight where experimentation cannot (at least at the present).

This chapter features some of the more recent major advancements in optical or visual measurements being made of combustion as used in internal combustion engines. I am gratefully indebted to the authors' whose work follows for their dedicated contribution to this chapter. Roughly half the submissions involve internal combustion engines of the reciprocating piston type, including those from Drs. Busch, Ciatti, Mueller, Musculus, and Sjoberg. The remaining half involve visualization of the combustion present in open internal combustion engines, such as jet engines and gas turbines, and include the contributions by Drs. Han, Kero, Stohr, and Versailles. What is most intriguing about these visualizations is not so much the qualitative features one can extract – which by themselves are incredibly useful particularly from an educator's perspective – but the quantitative features that are used to make impactful conclusions that change the course of research or execute substantial design decisions leading to a better engineered product.

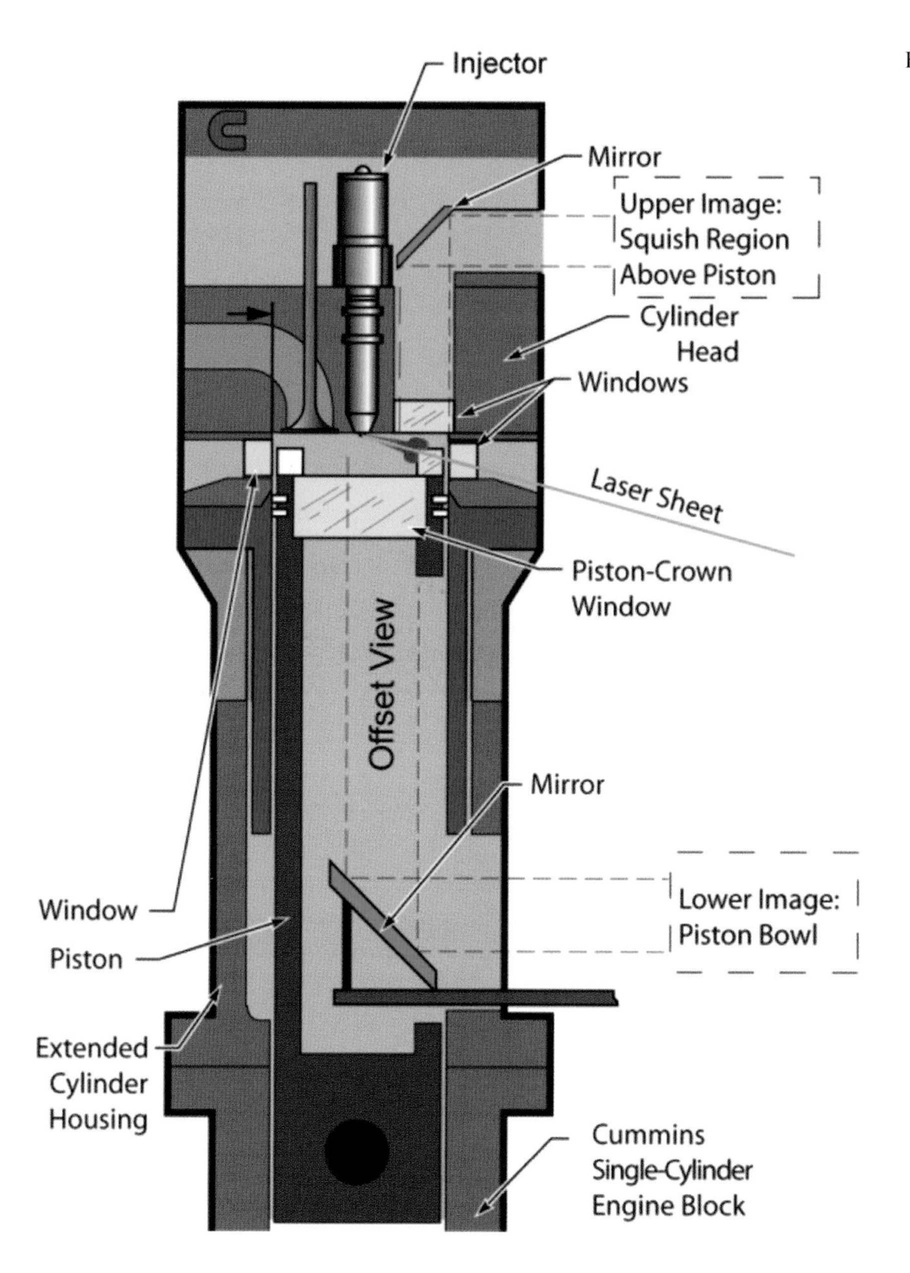

Figure 3.1

3.1 Single-Cylinder Version of a Cummins Six-Cylinder N-14 Highway Truck Engine

Mark Musculus

Sandia National Laboratories

Fuel injection, combustion, and pollutant formation processes are studied in an optically accessible, single-cylinder version of the Cummins six-cylinder N-14 highway truck engine. Windows in the cylinder walls provide access for laser illumination, while windows in the extended piston and cylinder head provide imaging access to the combustion bowl and "squish" region above the piston. Using lenses, high-energy pulsed laser beams are formed into thin sheets to probe the cross-section of the in-cylinder jets, which are imaged by intensified cameras.

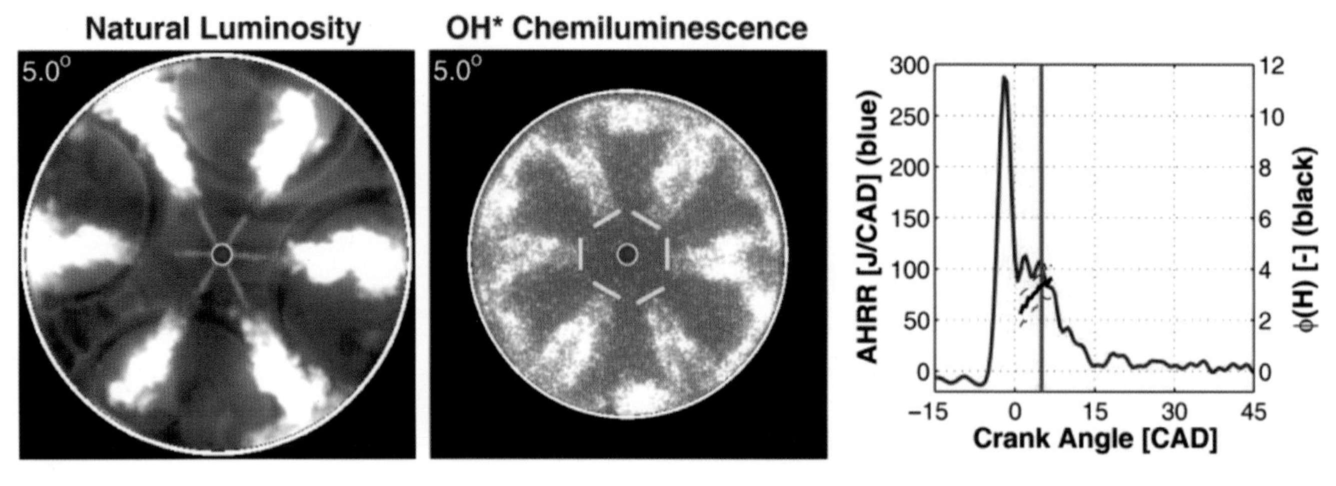

Figure 3.2

3.2 Mixing-Controlled Combustion in a Heavy-Duty Compression-Ignition Engine

Charles J. Mueller and Ryan K. Gehmlich

Sandia National Laboratories

This figure shows an in-cylinder image of natural luminosity (NL, left), an in-cylinder image of electronically excited hydroxyl radical chemiluminescence (OH* CL, center), and the apparent heat-release rate (AHRR) and the estimated equivalence ratio at the flame lift-off length ($\phi(H)$, right) during conventional, mixing-controlled combustion in a heavy-duty diesel engine. The images were acquired at 5.0 crank-angle degrees (CAD) after top-dead-center by viewing the combustion through a fused-silica window in the bowl of the piston. The piston bowl diameter is indicated by the larger white circle in each image, and the location of the fuel injector tip is indicated by the red dot at the center of each image. The NL signal is dominated by incandescence from hot soot, with the six saturated (white) regions corresponding to soot from each of the six fuel sprays. Liquid fuel in the six fuel sprays is also visible as gray lines emanating from the injector tip due to NL signal from hot soot being elastically scattered from the fuel droplets to the camera. The NL and OH* CL cameras detected photons with wavelengths in the range of 380–1000 nm and 308 ± 10 nm, respectively. The OH* CL image shows regions where high-temperature chemical reactions are occurring. The distance from the injector orifice exit to the most upstream extent of this region is known as the flame liftoff length, and the liftoff length for each spray is shown with a yellow bar. The estimated $\phi(H)$ values shown in the plot on the right are > 2, indicating that soot should be present, which is consistent with the hot soot evident in the NL image. The AHRR plot shows a typical premixed-burn spike followed by a period of mixing-controlled heat release, which is relatively brief at this 6 bar gross indicated mean effective pressure operating condition. Other engine operating conditions are: 1,500 rpm speed, 240 MPa injection pressure, −5 CAD start of combustion timing, 21% intake-oxygen mole fraction, 30°C intake manifold temperature, and 85°C coolant temperature. The red vertical line on the AHRR figure indicates the timing that corresponds to the images shown, which were acquired when 70% of the cumulative fuel mass was burned. The fuel is a commercial #2 diesel fuel.

Keywords:
Diesel; combustion; mixing-controlled; liftoff length

Reference

1. R. K. Gehmlich, C. E. Dumitrescu, Y. Wang, C. J. Mueller, Leaner lifted-flame combustion enabled by the use of an oxygenated fuel in an optical CI engine, *SAE International Journal of Engines* 9, 3 (2016), doi:10.4271/2016-01-0730.

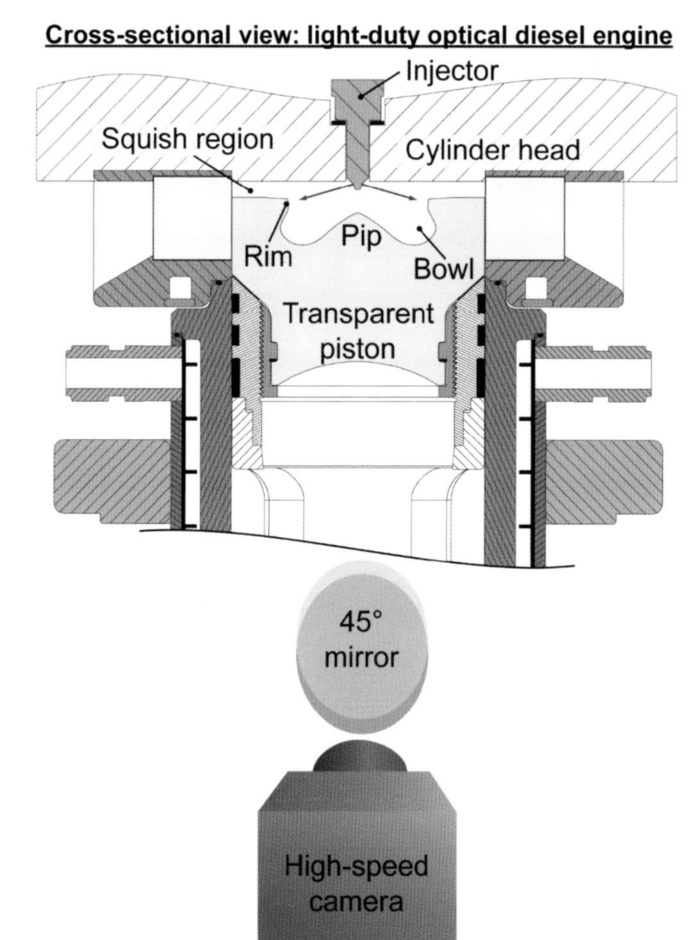

Figure 3.3

3.3 Visible Combustion Emissions in a Swirl-Supported, Light-Duty Diesel Engine

Stephen Busch

Sandia National Laboratories

This series of images taken with a high-speed camera show visible and near-infrared combustion emissions inside a swirl-supported, optical diesel research engine. The images are monochromatic, but are shown with a false color scaling to help distinguish regions of high intensity (yellow and white) from regions of low intensity (dark-red and orange). Images are taken through the engine's transparent piston, looking up at the cylinder head, as shown by the cross-sectional view. The thick white circle represents the edge of the field of view, which is several millimeters smaller than the cylinder bore.

Figure 3.4 (560 μs after the start of injection – ASOI): Seven jets of liquid fuel issuing from the centrally mounted fuel injector are dimly illuminated by the hot, glowing soot particles that form in regions where fuel has mixed with hot air and begun to burn. The swirling air in the cylinder has transported the fuel–air mixture so that the soot clouds are now displaced from the liquid jets in a clockwise direction. The presence of these glowing soot particles indicates hot combustion of rich mixtures. The piston is just starting to move downward at the beginning of the expansion stroke.

Figure 3.5 (760 μs ASOI): combustion continues as the outwardly propagating sprays impinge on the piston bowl rim. Part of the combusting mixture is deflected down into the piston bowl, and part of the mixture continues to propagate outward into the region between the piston and the cylinder head (this is called the squish region).

Figure 3.6 (960 μs ASOI): The injection of liquid fuel is nearing completion. Mixture in the bowl follows the contour of the wall and is redirected inward and upward along the central "pip" while also being transported by the clockwise swirl. This three-dimensional flow pattern is often referred to as a toroidal vortex. Outward propagation continues into the squish region.

Figure 3.7 (1,360 μs ASOI): The last fuel to be injected mixes poorly, and soot forms as it burns near the center of the cylinder. The toroidal vortex in the bowl supports the mixing of partially burned fuel and air, as the rate of fuel energy conversion to heat is near its maximum. Combusting mixture in the squish region continues to propagate outward.

Figure 3.8 (1,760 μs ASOI): Regions of hot soot are consumed by combustion reactions in the squish region. Soot intensities in the bowl are near their maximum values. The inward motion associated with the toroidal vortex in the bowl begins to decay, but swirling motion continues.

Figure 3.9 (3,320 μs ASOI): Late in the combustion event, most of the hot soot is observed in the bowl, and the downward motion of the piston acts to expand and cool the cylinder contents. Oxidation of soot requires high temperatures, as well as mixing and reaction

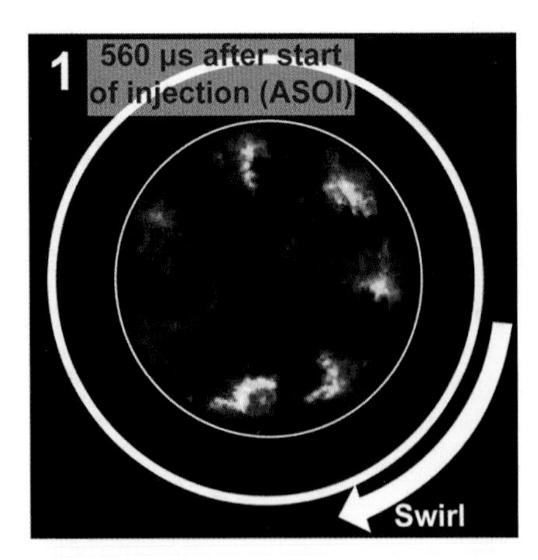

Figure 3.4

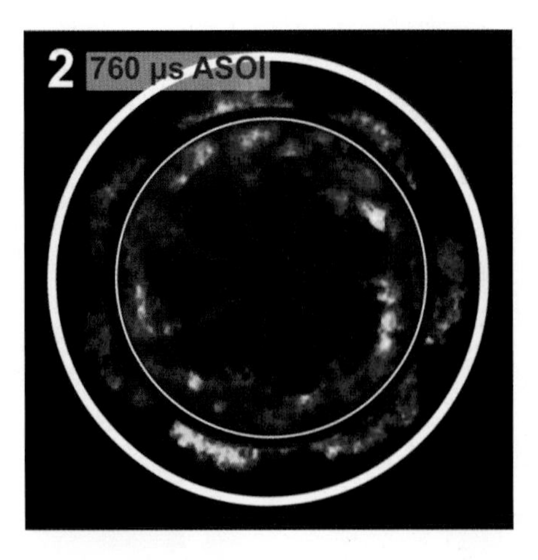

Figure 3.5

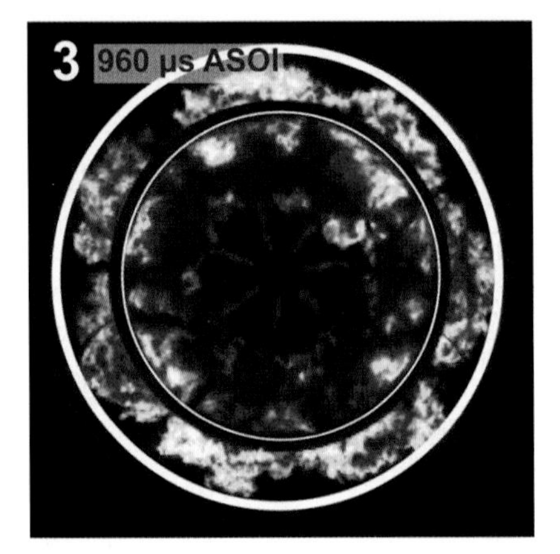

Figure 3.6

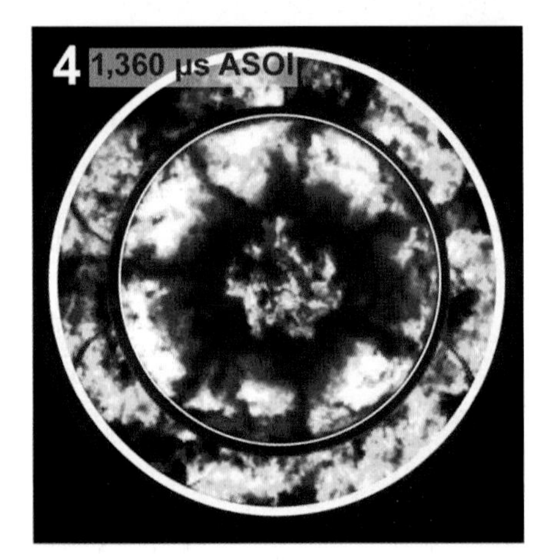

Figure 3.7

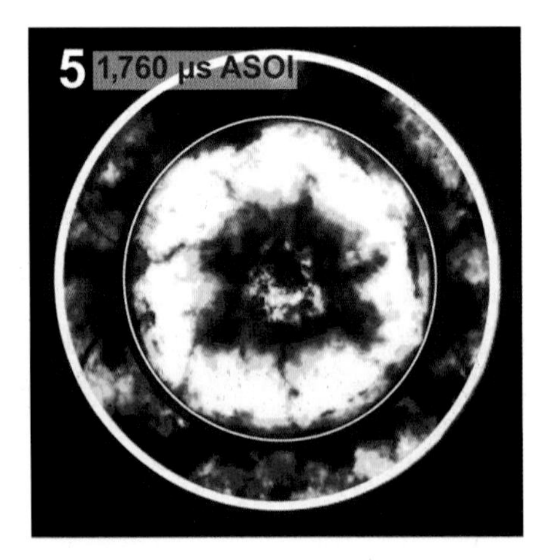

Figure 3.8

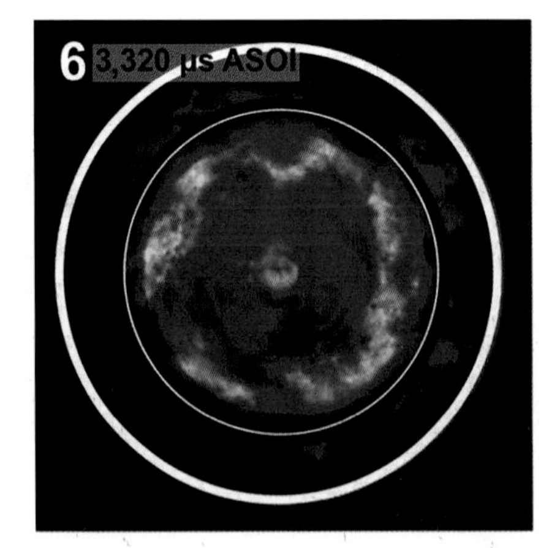

Figure 3.9

with an oxidizer, so turbulent mixing processes and chemical kinetics play important roles in this stage of combustion.

Keywords:
Diesel engine combustion; soot luminosity

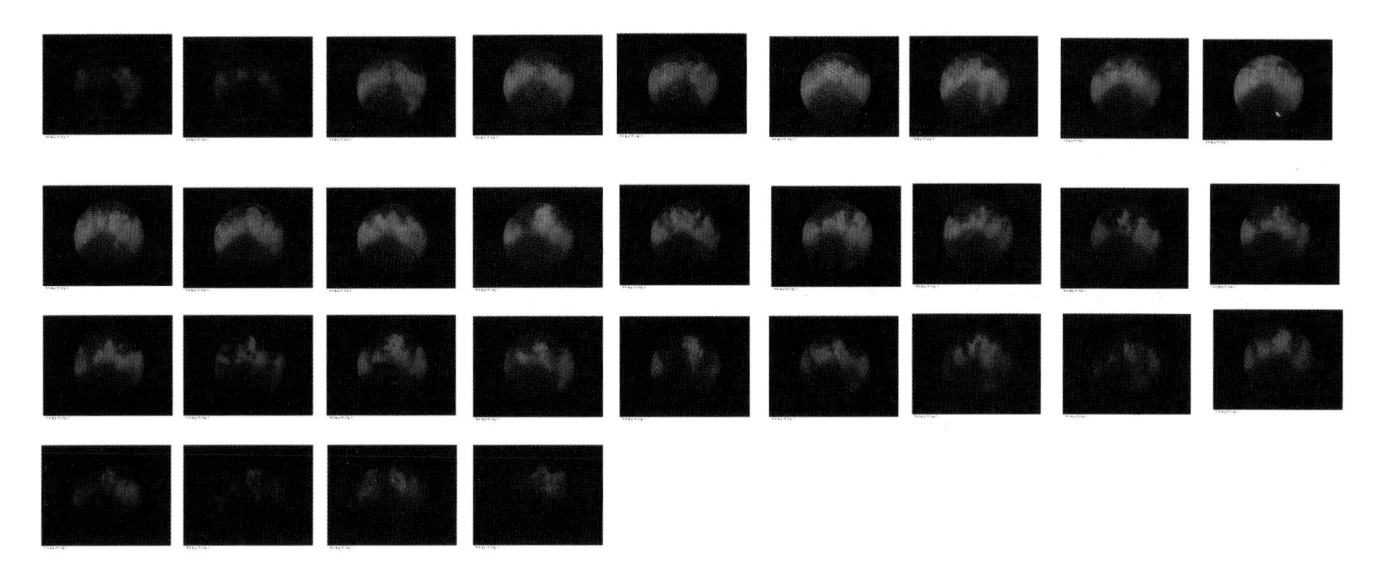

Figure 3.10

3.4 Sequential Images of Gasoline Compression Ignition Inside an Engine

Stephen Ciatti
Argonne National Laboratory

High-speed imaging captured this sequence of the auto-ignition and combustion progression of gasoline used in a compression ignition engine. The work is important to defining the compression ignition characteristics of gasoline rather than standard octane ratings empirically defined for spark ignition, stoichiometric engines. The images are compared with advanced engine simulations to better understand the mechanism of gasoline auto-ignition.

Keywords:
Gasoline compression ignition; low temperature combustion; advanced combustion

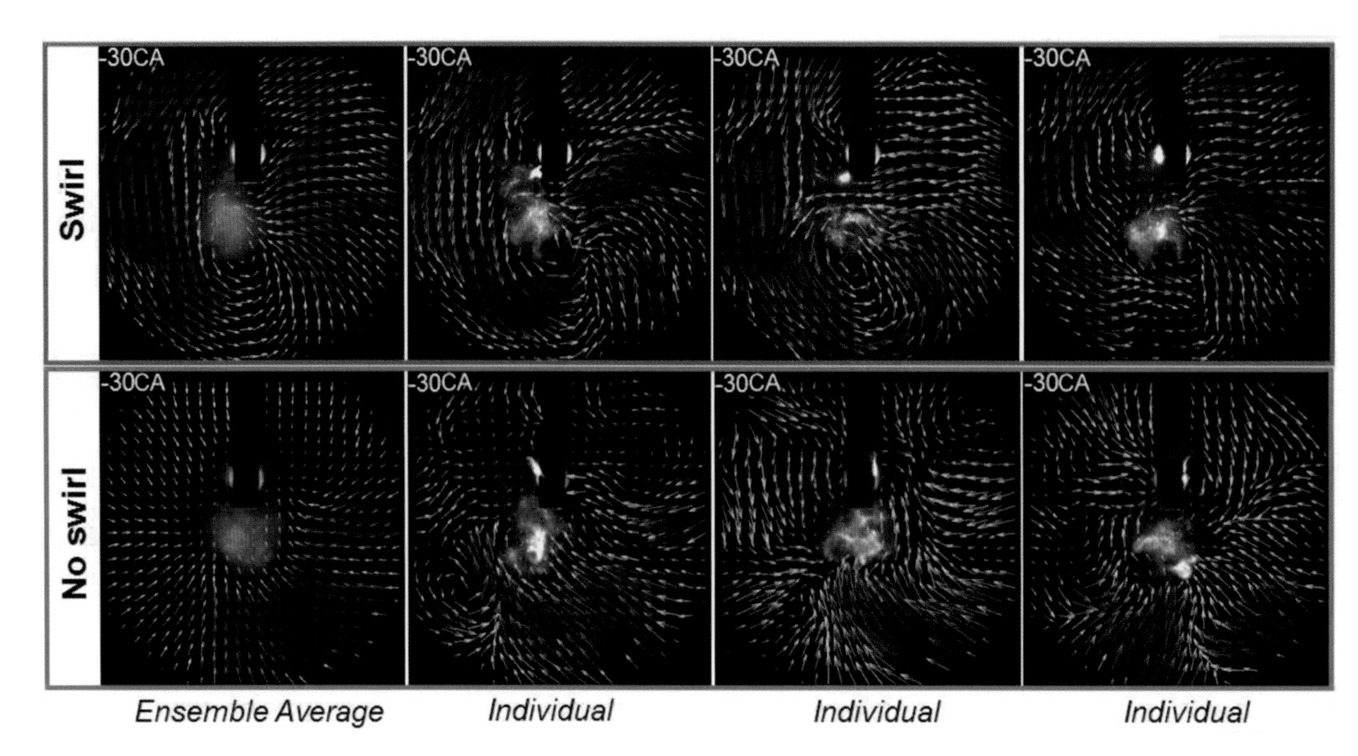

Figure 3.11 Cycle-to-cycle variations of flow pattern near spark plug 5°CA after EOI (end of injection)

3.5 Spray–Swirl Interactions Stabilize Stratified-Charge SI Operation by Reducing Flow Variability near the Spark

Magnus Sjöberg
Sandia National Laboratories

Research at Sandia National Laboratories has clarified the mechanism through which intake-generated swirl promotes stable combustion in stratified-charge spark-ignition (SI) engines operated on gasoline-type fuels. Using an optically accessible engine, the in-cylinder flow was examined with particle image velocimetry (PIV) for both swirling and non-swirling flows at an engine speed of 2,000 rpm. The measurements show that swirl makes the flow patterns of individual cycles more similar to the ensemble-averaged cycle, as exemplified in Figure 3.11. This effect is quantified in Figure 3.12 using a "flow similarity" parameter (R_P). With in-cylinder swirl, R_P remains consistently high for all examined cycles, indicating a high similarity between the ensemble-averaged flow and each individual cycle. Combustion data indicate a low IMEP variability of 1.4% for operation with swirl. In contrast, for

$$R_P = \frac{(D^{(1)}, D^{(2)})}{\|D^{(1)}\| \cdot \|D^{(2)}\|}$$

Relevance Index (R_p)
Stable flow pattern
Unstable flow pattern
2,000 rpm at 1.17 ms ASOI (-30 CA)
w/ swirl
w/o swirl
Cycle Number

Figure 3.12 Relevance index variations of flow patterns with and without swirl

operation without swirl, the average R_P is lower, and some cycles have very low R_P values. This leads to an unacceptably high IMEP varibility of 3.5%.

The interaction of the eight fuel sprays with the swirling gas flow strongly contributes to the flow stabilization at the time and location of the spark, as described

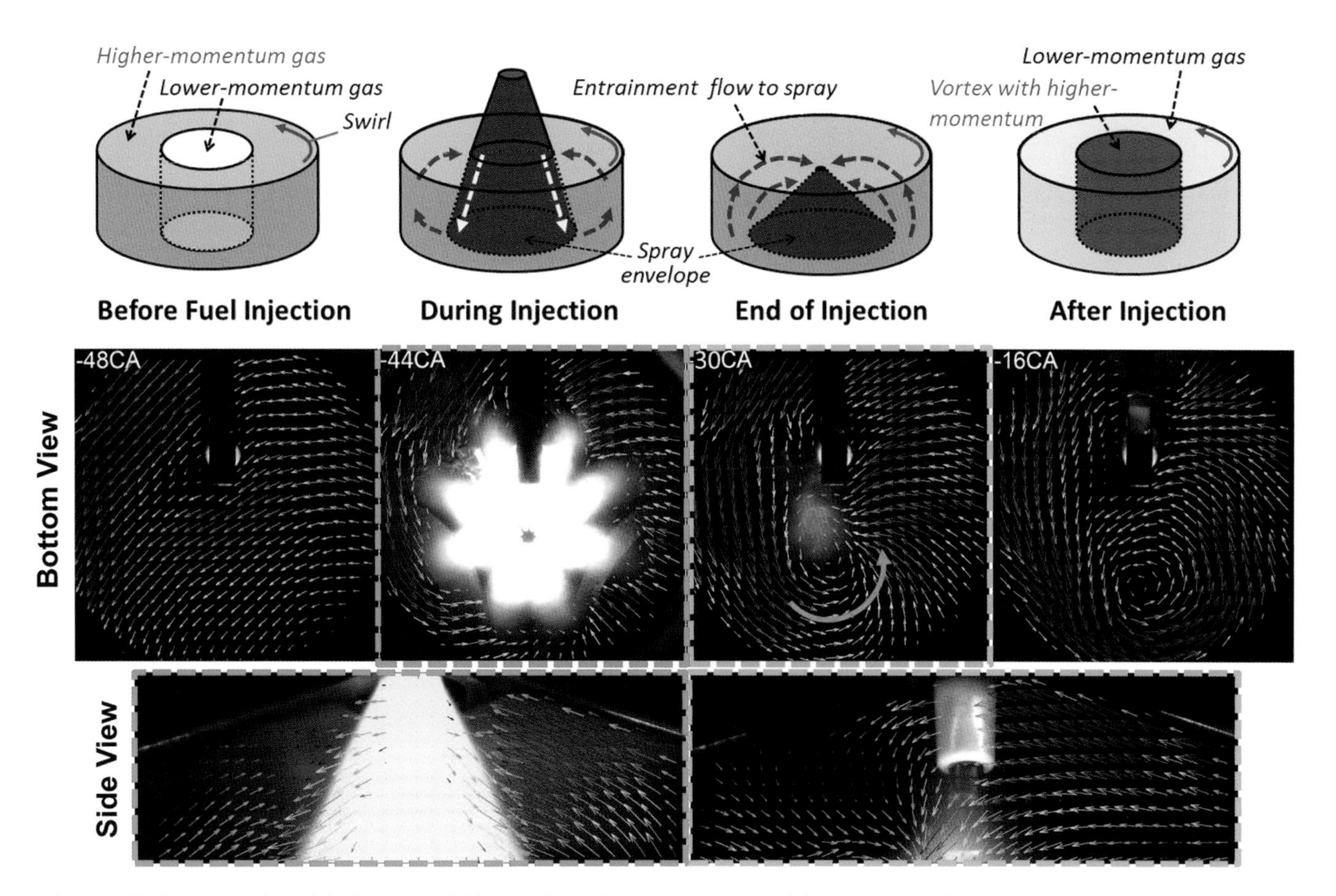

Figure 3.13 Conceptual model of spray-swirl interactions that create a repeatable vortex near the spark plug, based on PIV imaging experiments

conceptually in the top portion of Figure 3.13. The flow measurements reveal that the fuel sprays displace low angular-momentum gas downward. In the wake of the spray, higher-angular-momentum gas from larger radii flows inward and increases its rotation rate due to conservation of momentum. This process creates a strong and very repeatable vortex near the spray centerline at the time of spark, as demonstrated by the vector fields in Figure 3.13. Related research efforts have shown that the increased flow similarity stabilizes the combustion process by reducing variability in both the ignition event and in the flame spread throughout the remaining charge; see Ref. 1. This avoids the occurrence of slow-burning cycles that may develop into partial burns. In contrast, the flow fields without swirl in the bottom row of Figure 3.11 show no evidence of such a stabilizing vortex. Moreover, the single-cycle examples without swirl are quite different from the average flow field, consistent with the observed higher combustion variability. Using this understanding of swirl–spray interactions, combustion engineers can optimize injector and flow parameters to maximize combustion stability and enable highly efficient stratified-charge SI operation.

Keywords:
Stratified direct-injection spark-ignition engine; cycle-to-cycle variations; combustion stabilization

References

The top portion of Figure 3.13 was published previously in:

W. Zeng, M. Sjöberg, D. L. Reuss, PIV examination of spray-enhanced swirl flow for combustion stabilization in a spray-guided stratified-charge DISI engine, *International Journal of Engine Research* 16, 3 (2014), 306–322.

An appropriate reference for the discussed stabilization of the flame spread is:

1. W. Zeng, M. Sjöberg, D. L. Reuss, Z. Hu, The role of spray-enhanced swirl flow for combustion stabilization in a stratified-charge DISI engine, *Combustion and Flame* 168 (2016), 166–185. http://dx.doi.org/10.1016/j.combustflame.2016.03.015

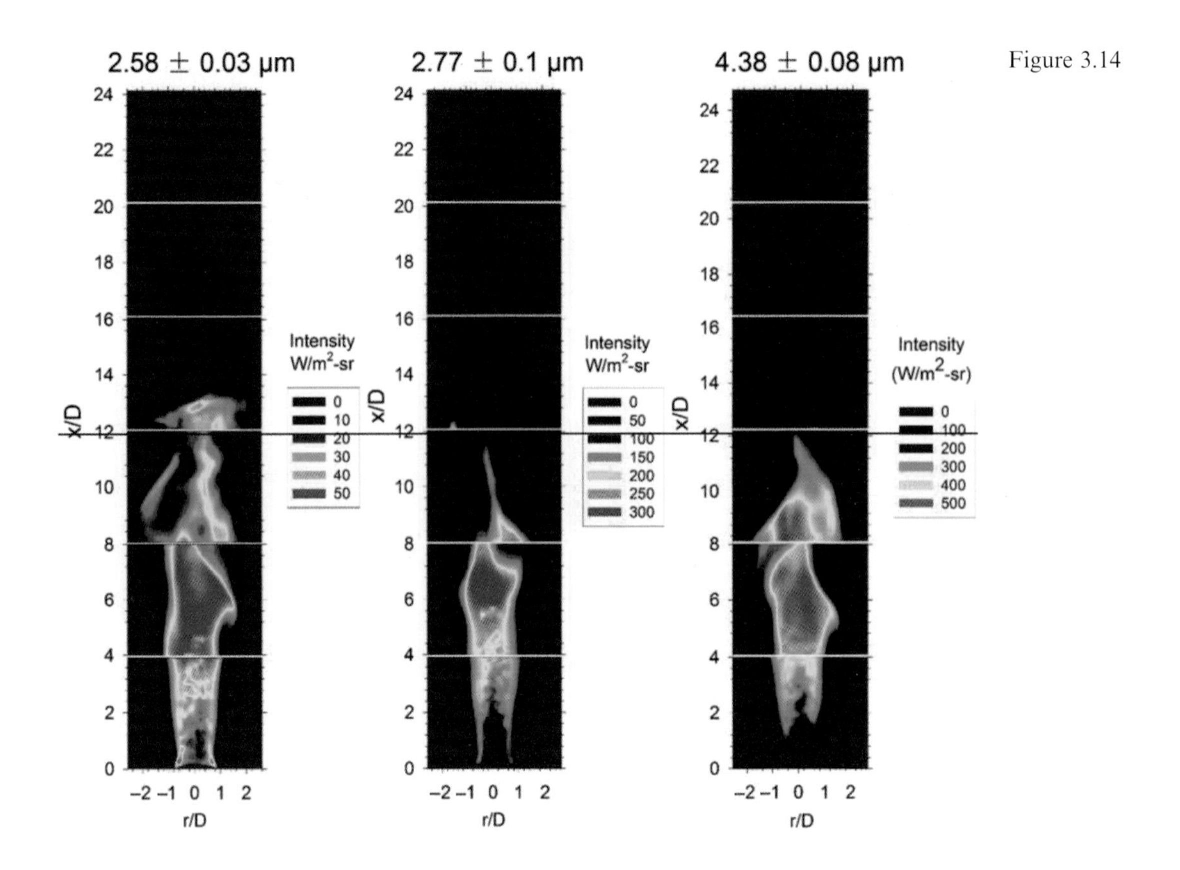

Figure 3.14

3.6 Quantitative Narrowband Infrared Imaging of a Turbulent Premixed Flame

Dong Han and Jay Gore

Purdue University

Figure 3.14 shows quantitative time-dependent measurements of the narrowband (2.58 ± 0.03 μm, 2.77 ± 0.1 μm, and 4.38 ± 0.08 μm) radiation intensities emitted from a turbulent premixed flame. Figure 3.15 shows ensemble averaged images of 6,400 time-dependent images for the three infrared narrow bands. The radiation intensity measurements of the flame in the 1–3 μm range were acquired using a FLIR Phoenix camera (InSb detector array with liquid nitrogen (LN_2) cooling) and those in the 3–5 μm range were acquired using a FLIR SC6000 camera (InSb detector array with thermoelectric (TE) cooling). The images were taken at 430 Hz with an exposure time of 10 μs. Each image consists of six panels. These panels are temporally independent but spatially correlated. Figure 3.16 shows a comparison of measured and computed radiation intensity of 4.38 ± 0.08 μm band emitted from the turbulent premixed flame. Measured temperature profiles, state relationship, and stochastic time and space analysis modeling turbulent radiation interaction were employed for radiation intensity computation. Emission coefficients were calculated using narrowband model from RADCAL or line-by-line model with line strengths and widths from HITRAN and HITEMP databases.

The flame was established on a rim stabilized co-flow burner with an inner diameter (D) of 15 mm. A methane flow rate of 84 ± 1 mg/s and an equivalence ratio of 0.8 were chosen for the present measurements. A pilot flow of hydrogen at 2 ± 0.05 mg/s through a ring of annular holes was applied to stabilize the flame. The nominal jet exit Reynolds number (8950 ± 50) is based on the cold gas properties, the exit velocity of the central jet burner, and the burner diameter.

This combined experimental and computational work studied infrared radiation intensity emitted by individual and combined bands of H_2O and CO_2 from a turbulent premixed flame. The purpose was to study the thermal radiation properties of turbulent premixed

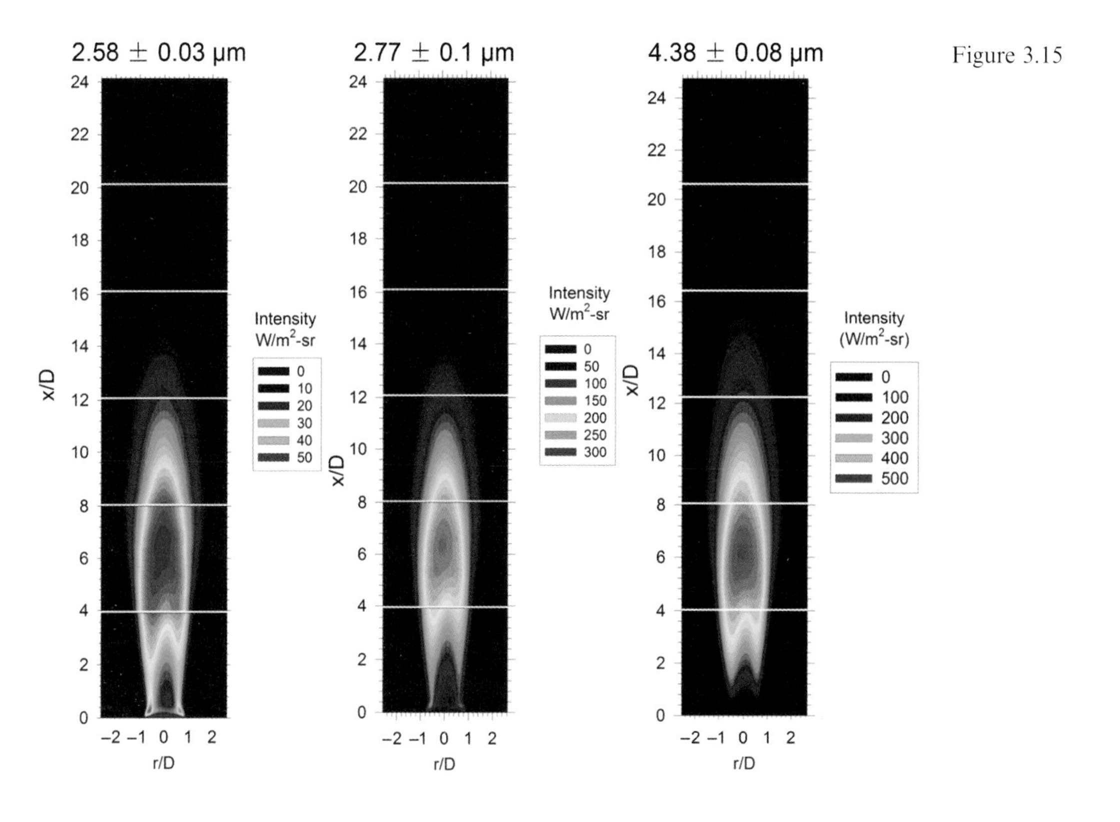

Figure 3.15

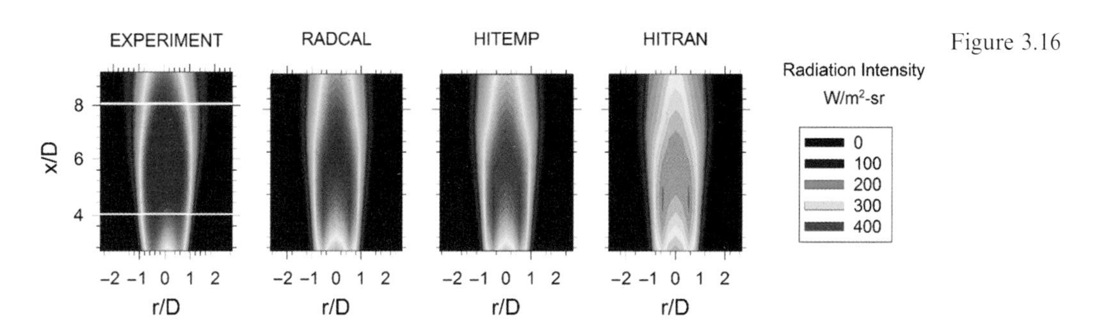

Figure 3.16

flames, which provides insight into flame structures by revealing detailed features of the flame stabilization zone near the burner exit, the wrinkled flame zone at intermediate distances, and the broader combustion zone near the flame tip. In addition, the agreement between measured and computed radiation intensities of a turbulent premixed flame demonstrates a potential application in full-scale industrial facilities such as gas turbine combustors for combustion diagnostics and computational model validation.

This work was supported by the US Department of Energy (DOE), National Energy Technology Laboratory (NETL), University Turbine Systems Research (UTSR) Program with DOE award number DE-FE0011822.

Keywords:

Turbulent premixed flames; infrared radiation imaging; stochastic analysis; turbulent radiation interaction

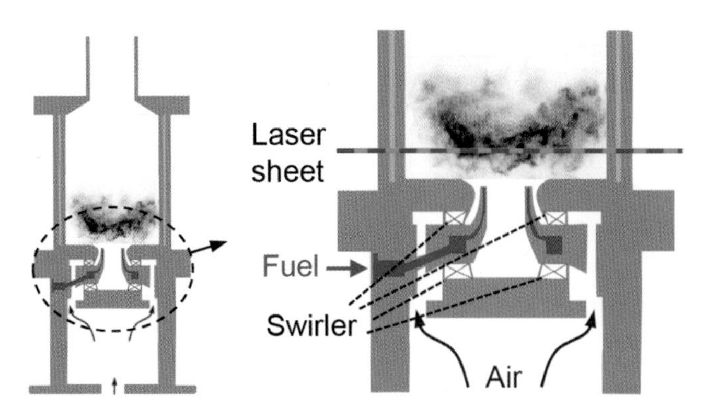

Figure 3.17

3.7 Thermoacoustic Oscillation of a Confined Turbulent Swirl Flame

Michael Stöhr and Wolfgang Meier
German Aerospace Center (DLR)

Rajesh Sadanandan
Indian Institute of Space Science and Technology (IIST)

Turbulent swirl flames are commonly employed to drive gas turbines in stationary power plants or aeroengines. A major technical challenge is the occurrence of self-excited, so-called thermoacoustic instabilities that arise due to a complex feedback mechanism involving interactions of flow field, acoustics, and heat release. The instabilities lead to strong pressure oscillations that affect combustor performance and lifetime.

The feedback mechanisms have been studied experimentally in a partially premixed model gas turbine combustor with optical access. The combustor is operated with methane and air at a thermal power of 10 kW and an equivalence ratio of 0.75, where a thermoacoustic oscillation of ≈300 Hz occurs. Instantaneous flow fields and reaction zones have been measured simultaneously in a transverse section using stereoscopic particle image velocimetry and planar laser-induced fluorescence of the OH radical, respectively.

The images show combinations of OH distribution (gray scale) and velocity field (colored vectors). Regions without OH (black) represent unburned gas, whereas high levels of OH (light-gray to white) indicate superequilibrium OH formed in the reaction zones. Medium and low levels of OH (medium and dark gray) represent burned gas whose OH concentration has decayed toward equilibrium while it

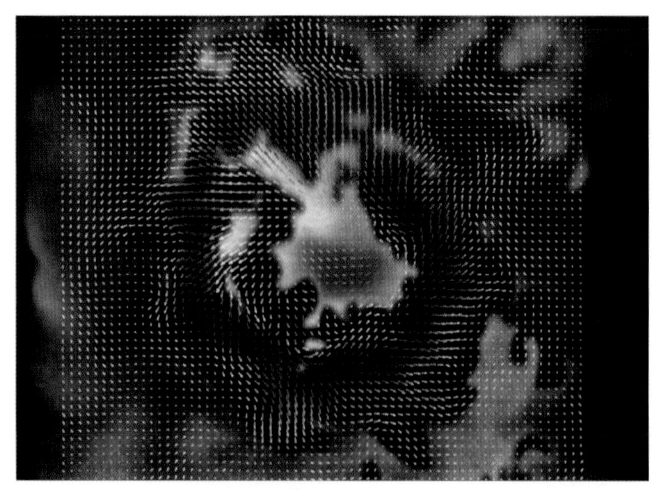

Figure 3.18

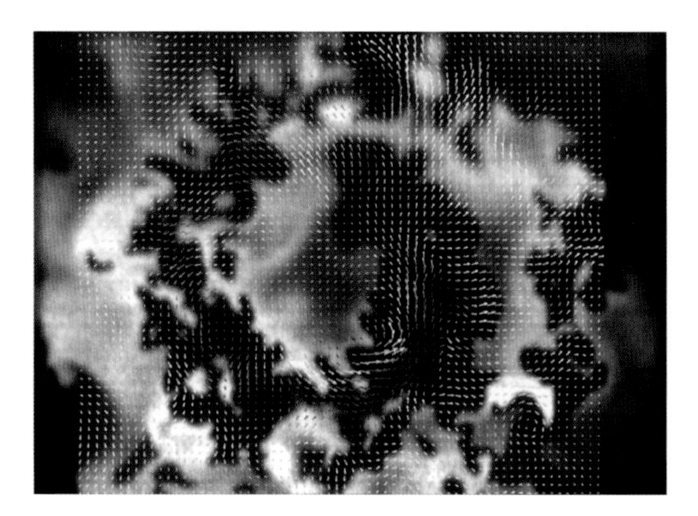

Figure 3.19

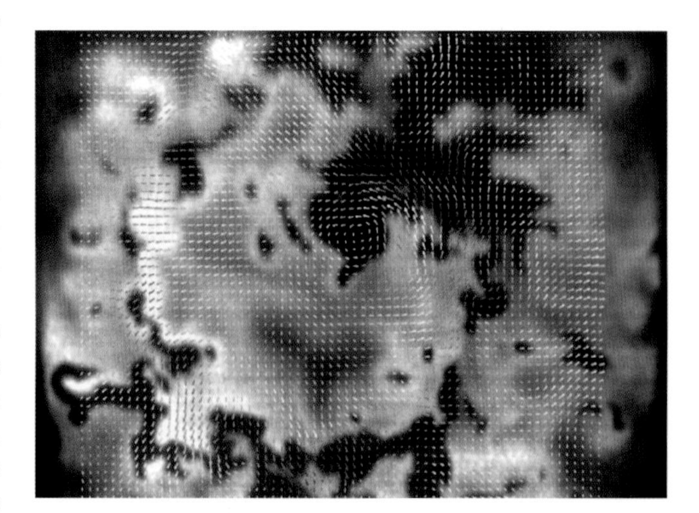

Figure 3.20

was transported away from the reaction zone. The vector color indicates the out-of-plane velocity.

The three images represent different times during the cycle of the pressure oscillation. In the first image a strong inflow of unburned gas is observed (region of red vectors) and the reaction is confined to the inner zone. In the second image, most of the unburned gas burns along extensive wrinkled reaction zones, which leads to a strong overall heat release. As a result, burned gas is present in most part of the third image, and the overall heat release is reduced until the inflow of fresh gas at the beginning of the next oscillation cycle. In summary, the measurements reveal the considerable changes of the reaction zone topology associated with the combustion instability.

The work was funded by German Aerospace Center (DLR) within the project MVS.

Keywords:
Turbulent swirl flame; partially premixed flame; laser diagnostics; thermoacoustic instability

Reference

Similar images were published in:

M. Stöhr, R. Sadanandan, W. Meier, Experimental study of unsteady flame structures of an oscillating swirl flame in a gas turbine model combustor, *Proceedings of the Combustion Institute* 32 (2009), 2925–2932.

Figure 3.21

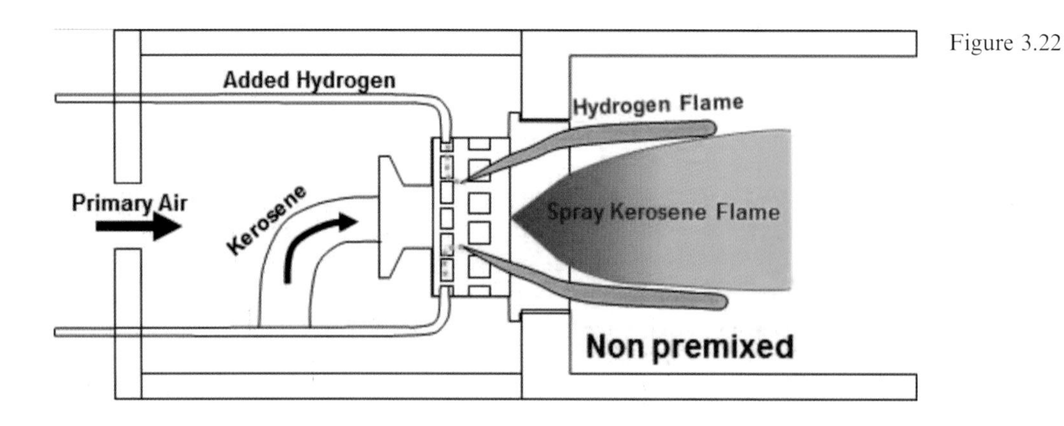

Figure 3.22

3.8 Effects of H_2 Enrichment on Flame Stability and Pollutant Emissions for a Kerosene/Air Swirled Flame with an Aeronautical Injector

Joseph Burguburu and Michel Cazalens
SAFRAN

Abdelkrim Mourad Boukhalfa, Bruno Renou, and Gilles Cabot
CORIA, INSA

Experimental studies are conducted on an aeronautical kerosene spray injection system to characterize the flame stability and pollutant emissions of two-phase kerosene/air flames, at conditions similar to the idle phase of an aircraft engine (P = 0.03 MPa and T = 500 K). A multi-point fuel injector (SAM 146 aircraft engine) is provided by SAFRAN-AE and integrated in a fully transparent combustion chamber (Figure 3.21). Hydrogen as a surrogate of reformer gases from kerosene reforming is added in a small amount to enhance flame stability and to reduce NO_x and CO on the idle and taxi phases. Hydrogen is injected directly into three vanes (120 degrees) of the central air swirler of the injector by three pipes, with an external diameter of 3 mm (Figure 3.22). When hydrogen is used and the kerosene is off, one can see three hydrogen/air pilot flames at the exit of the injector in the bottom of

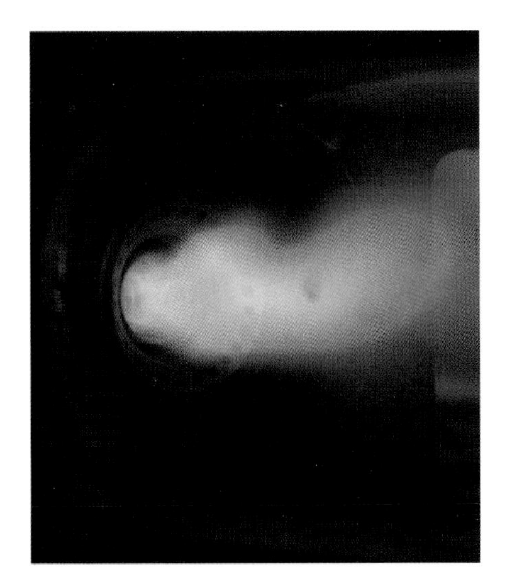

Figure 3.23

the combustion chamber (Figure 3.23). When the kerosene is on, a classical swirled flame is obtained (Figure 3.24) and the flame stability is improved, and the lean blow-off limit is reduced by the presence of these three pilot flames.

The injection system consists of two air swirlers and two kerosene-supplying pipes. Considering the operating condition (1/15 of the takeoff setting), only the first kerosene-supplying pipe is used. The air blast injector generates a spray SMD close to 50 microns. The swirl number is 0.73, with the air and kerosene mass flow rates of 35 g/s and 1.6 g/s, respectively. The amount of H_2 enrichment is 2.8%, defined by the ratio of the H_2 to kerosene mass flow rate corrected by the ratio of their low heating values.

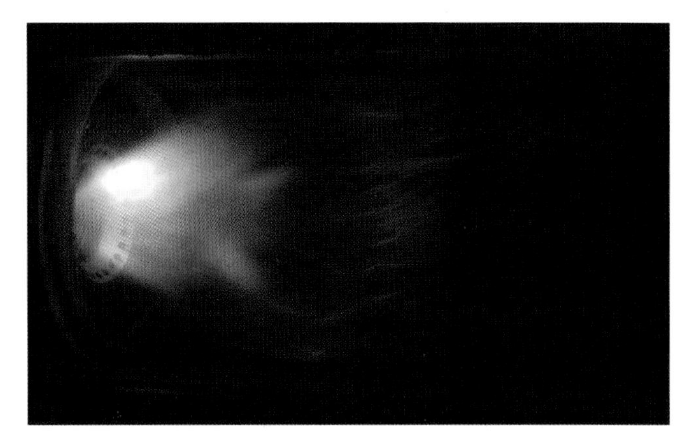

Figure 3.24

The authors gratefully acknowledge the financial support of SAFRAN-AE and SAFRAN Tech.

Keywords:
Aeronautical kerosene injector; H_2 fuel enrichment; flame stability

References

1. J. Burguburu, G. Cabot, B. Renou, A. Boukhalfa, M. Cazalens, Effects of H_2 enrichment on flame stability and pollutant emissions for a kerosene/air swirled flame with an aeronautical injector, *Proceedings of the Combustion Institute* 33, 2 (2011), 2927–2935.
2. J. Burguburu, G. Cabot, B. Renou, A. Boukhalfa, M. Cazalens, Comparisons of the impact of reformer gas and hydrogen enrichment on flame stability and pollutant emissions for a kerosene/air swirled flame with an aeronautical fuel injector, *International Journal of Hydrogen Energy* 36, 11 (2011), 6925–6936.

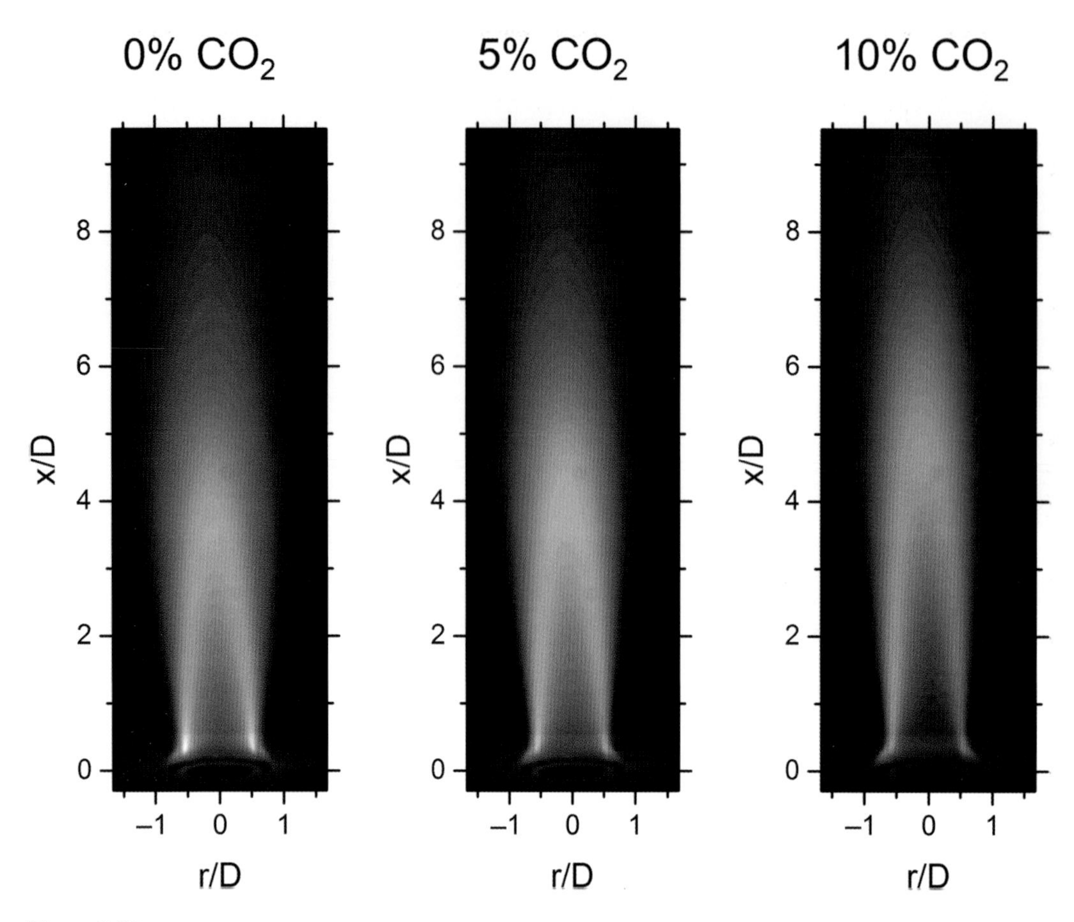

Figure 3.25

3.9 Effects of Carbon Dioxide Addition on Turbulent Premixed Flames

Dong Han, Aman Satija, Jay Gore, and Robert Lucht

Purdue University

Figure 3.25 shows averaged visible images of turbulent premixed flames with varying CO_2 addition of 0%, 5%, and 10%. Each image is an ensemble average of 100 images taken by a Phantom v7.1 at 100 Hz with 10 ms exposure time. Figure 3.26 shows instantaneous images using OH laser induced fluorescence (PLIF) for these flames. The image was taken by a Phantom v7.3 camera with a LaVision high-speed IRO intensifier at 9 kHz with an exposure time of 50 ns. Each image consists of nine panels. These panels are temporally independent but spatially correlated. Each panel has a 15% overlap with an adjacent panel along the axial axis of the burner centerline. The bright areas in the images indicate combustion products with a finite concentration of the OH radical, while the dark areas show unburned reactants with no detectable OH radical. The brightness of each pixel shows the OH fluorescence intensity received by the camera.

The flames were established on a piloted axisymmetric reactor assisted turbulent (PARAT) burner. The burner nozzle exit has an 18 mm diameter (D) annular cross-section with a 3 mm-thick wall. The burner is 523 mm tall with diverging (bottom) and converging (near the nozzle exit) sections. To sustain a flame at large Reynolds numbers, a non-premixed pilot flame of undiluted hydrogen was utilized. The pilot flame was generated by flowing hydrogen through a 2 mm-wide annulus just outside of the nozzle outer wall. Turbulence generator plates were installed within the burner. The lower turbulence generator plate is located before the final convergence of the flow channel, 177 mm upstream of the nozzle exit. This plate has five circular holes with a diameter of 6.35 mm and a 94% blockage ratio. The upper turbulence generator plate is placed 30 mm upstream

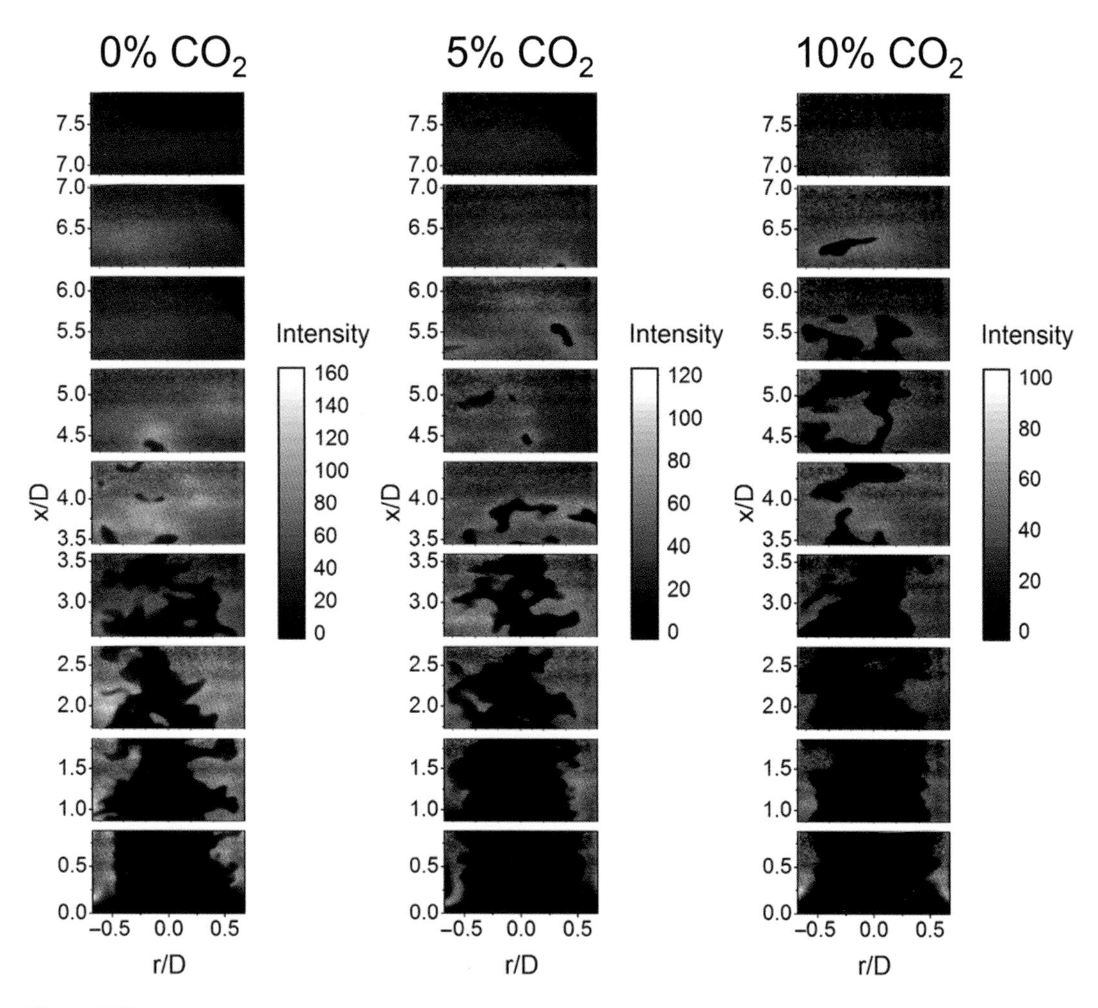

Figure 3.26

of the nozzle exit. The upper plate has 13 circular holes with a diameter of 2.54 mm and an 87% blockage ratio.

The flames were operated at identical jet Reynolds numbers (10,000), and the equivalence ratios were chosen as 0.8, 0.84, and 0.89 for flames with, respectively, 0%, 5%, and 10% CO_2 addition to yield identical adiabatic flame temperature (2,030 K) in order to minimize the thermal effects on flame propagation and structure. The unburned mixture Lewis numbers of the flames differed by a maximum of 2%, thereby minimizing transport effects on the flame structure. The heat release rates for the turbulent premixed flames and pilot flames calculated using lower heat values were about 6 kW and 0.4 kW, respectively.

This experiment studied the effects of exhaust gas recirculation (EGR) emulated by CO_2 addition on turbulent premixed flames with minimized thermal and transport effects. The purpose was to better understand the effects of EGR on turbulent premixed combustion in gas turbine systems with identical turbine inlet temperature and generate an experimental database for the development and validation of a combustion model and a computational tool for gas turbine applications.

Funding for this work was provided by the US Department of Energy, National Energy Technology Laboratory (NETL), University Turbine Systems Research (UTSR) Program under Grant No. DE-FE0011822 and by the US Department of Energy, Division of Chemical Sciences, Geosciences, and Biosciences under Grant No. DE-FG02–03ER15391.

Keywords:

Turbulent premixed flames; CO_2 addition; OH PLIF

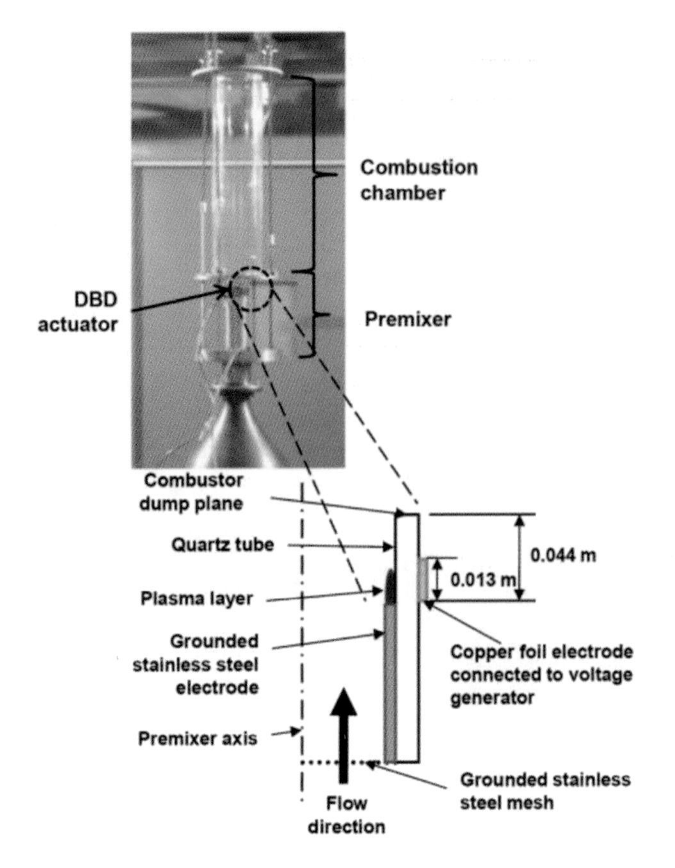

Figure 3.27

3.10 Application of a Dielectric Barrier Discharge (DBD) to Retard Flashback in Lean Premixed Dump Combustors

Philippe Versailles
McGill University

Wajid Ali Chishty
National Research Council Canada

Huu Duc Vo
École Polytechnique de Montréal

Lean, premixed combustors have the potential to significantly reduce nitrogen oxide emissions of gas turbine engines. However, as the fuel and air streams are mixed upstream of the combustion chamber, flashback can occur and significantly damage the premixer that is not designed to sustain high temperatures.

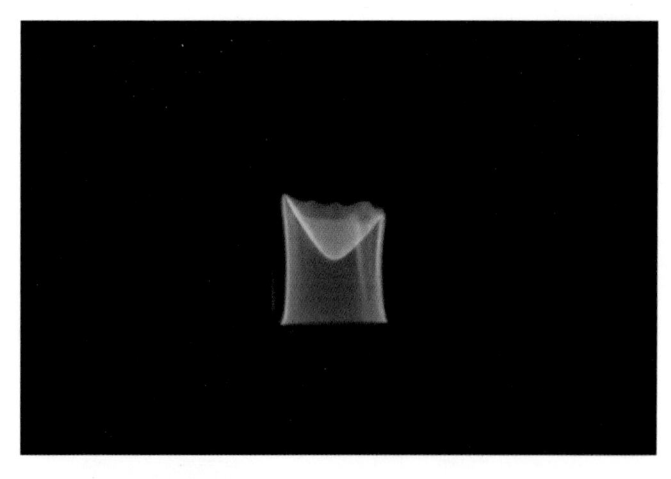

Figure 3.28

(a)

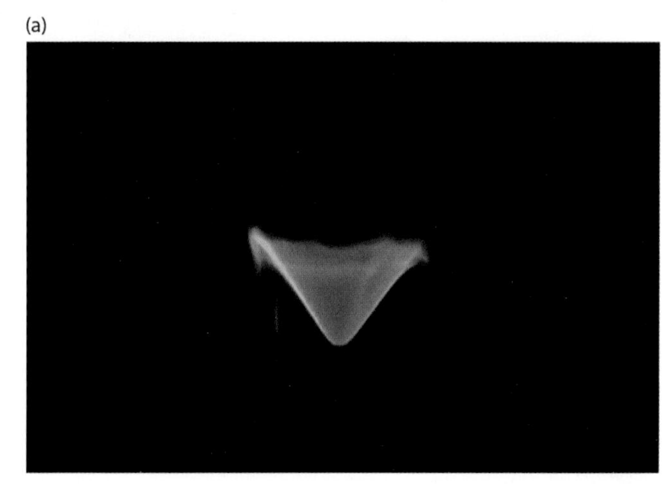

(b)

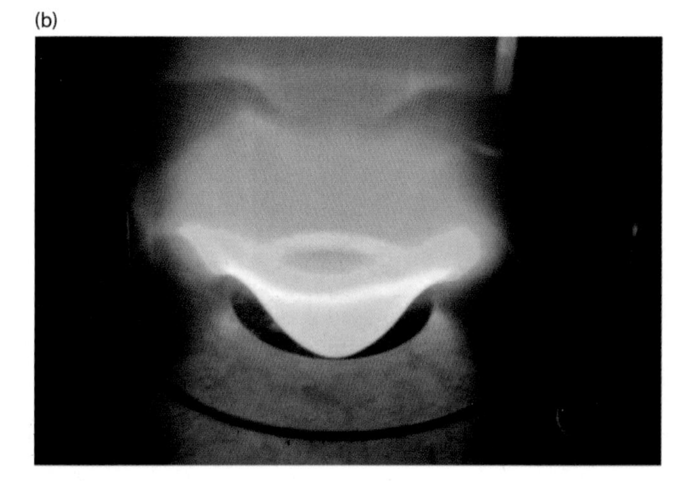

Figure 3.29

A technology based on non-thermal plasma discharges was developed at the National Research Council Canada. As shown in Figure 3.27, it employs a DBD actuator installed on the wall of the quartz premixer to produce a

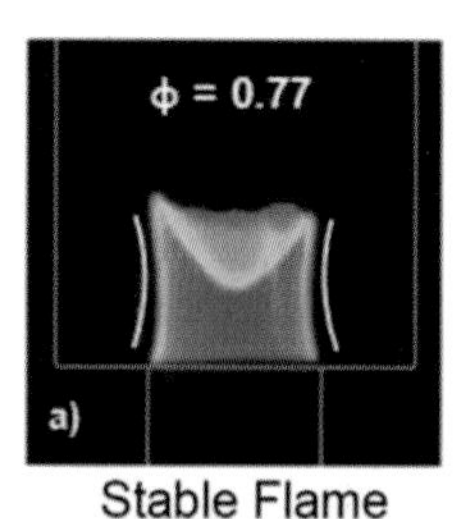

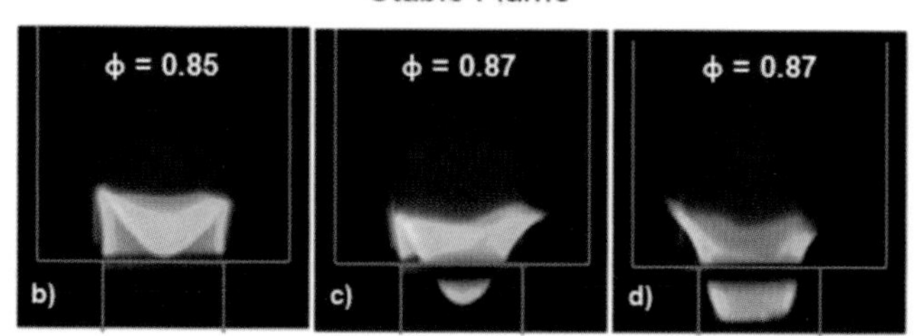

Figure 3.30

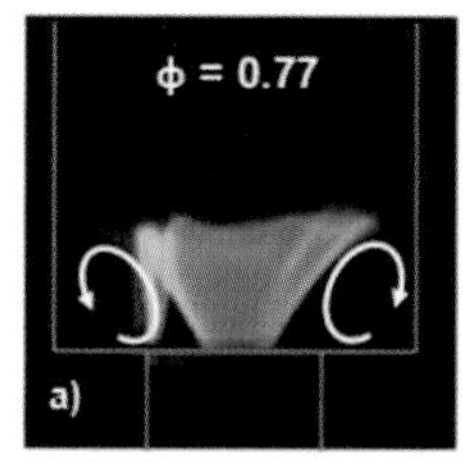

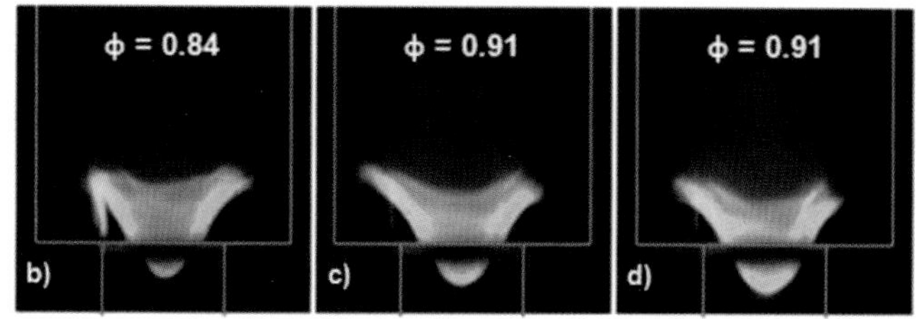

Figure 3.31

plasma layer. The charged molecules are entrained by the electric field across the electrodes and transfer their momentum to the neighboring neutral molecules. This results in a volumetric force close to the wall of the premixer, which locally accelerates the flow toward the combustion chamber. As the high-voltage, high-frequency AC electric signal is applied to the electrodes, the lean (equivalence ratio, φ, of 0.75) methane/air flame that was originally attached to the burner rim (Figure 3.28) is lifted and adopts a new stable position on the toroidal vortex located in the corner of the combustion chamber (φ = 0.75 in Figure 3.29(a) and φ = 0.99 in Figure 3.292(b)).

Figures 3.30 and 3.31 present a series of flame images taken with decreasing air flow rates for a fixed fuel flow rate of 0.102 g/s, without and with DBD actuation, respectively. The resulting reduction in overall flow rate and increase in equivalence ratio, leading to a rise in laminar flame speed, are two factors favoring flashback. It is observed that the actuated flame flashes back at a higher equivalence ratio, therefore demonstrating the ability of DBD actuators to prevent this phenomenon by changing the stabilization mechanism of flames in dump combustors.

This work was supported by the National Research Council Canada and the Natural Sciences and Engineering Research Council of Canada whose contributions are gratefully acknowledged.

Keywords:
Flashback; flow control; lean premixed combustor; non-thermal plasma discharge; plasma actuator

References

1. P. Versailles, W. A. Chishty, H. D. Vo, Application of dielectric barrier discharge to improve the flashback limit of a lean premixed dump combustor, *Journal of Engineering Gas Turbines Power* 134, 3 (2012), 031501-1–031501-8.
2. P. Versailles, W. A. Chishty, H. D. Vo, Plasma actuation control of boundary layer flashback in lean premixed combustor. Proceedings of the ASME Turbo Expo 2012, Copenhagen, Denmark, Paper GT2012–68224, 2012.

4 Low-Gravity Flames

Edited by Sandra Olson

Introduction

Sandra Olson

Our perception of a flame is strongly grounded in gravity's influence. From our every interaction with fire from the first birthday candles we blew out, we each build an intuitive understanding of how a flame interacts with the hot air rising via buoyant convection. As researchers, our perceptions of how flames respond to our controls are unconsciously biased by this intrinsic buoyant flow.

It has been my experience that many renowned researchers come into the microgravity field of study with preconceived notions of how their flames will respond when gravity is removed, only to be flummoxed when something completely unexpected happens. When gravity is removed, suddenly new realms of fluid mechanics, heat transfer, and chemical reactions become accessible, which is one of the really rewarding aspects of studying microgravity combustion. This is a space-age, still evolving field that is rich for study. It also does not hurt that some of the experimental platforms for microgravity combustion are literally out of this world.

Microgravity combustion does have its challenges with diagnostics, however. Diagnostics have to be small and low power due to launch weight and cooling constraints for flight experiments, and for some drop tower tests they must also be able to withstand repeated 65 g impacts and vacuum conditions. Due to these serious constraints, microgravity flame diagnostics have lagged behind laboratory diagnostic systems. One can imagine the laser alignment issues associated with experiment packages that flex as gravity is removed and flex again during the high-gravity impacts. Recent advances in digital cameras have revolutionized flame imaging so that the dimmest blue flames can be captured even with high-speed movies. Many previously "invisible" microgravity flames can now be imaged, such as cool flames.

This section is organized by general topic, with subsections noted in parentheses: gaseous fuel flames (spherical and gas jets), liquid fuel flames (droplet, candles, and pool fires), and solid fuel flames (spherical, rods, and thin and thick flat sheets).

For further study of microgravity combustion, a good resource is *Microgravity Combustion: Fire in Free Fall*, by Howard D. Ross, ed., Academic Press, 2001. It provides an excellent framework of understanding and a number of color plates of older microgravity combustion graphics.

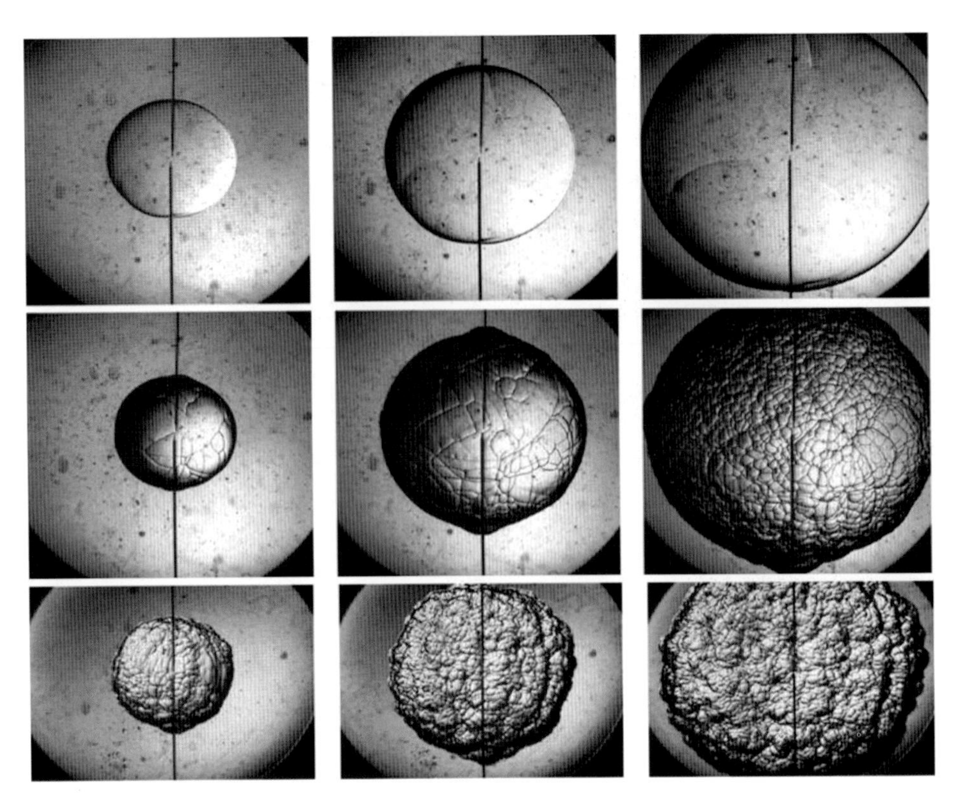

Figure 4.1 High-speed Schlieren images for spark-ignited spherically expanding flame under three different conditions demonstrating cellular instabilities that developed: **(top)**: Rich $H_2/O_2/N_2$ mixtures, equivalence ratio $\varphi = 1.5$, $P = 1$ atm, Le > 1; **(middle)** Rich $H_2/O_2/N_2$ mixtures, $\varphi = 1.5$, $P = 5$ atm, Le > 1; **(bottom)** Lean $H_2/O_2/N_2$ mixtures, $\varphi = 0.4$, $P = 2$ atm, Le < 1.

4.1 Gaseous Fuels

4.1.1 Cellular Instabilities in Premixed Flames

Abhishek Saha and Chung K. Law

Princeton University

For many combustible mixtures and ambient conditions, instead of a smooth flame surface (Figure 4.1 top), irregular cells can spontaneously develop and grow over the flame surface as the premixed flame propagates outward, as shown in Figure 4.1 middle and bottom.

The presence of these cells increases the total (wrinkled) flame surface area and consequently the global flame propagation speed. Furthermore, with the continuous generation of these cells, the increase in the propagation speed can be accelerative, leading to the hypothesized scenarios that a wrinkled laminar flame can first transition to a turbulent flame, and eventually to a detonation wave.

There are two instability mechanisms that are intrinsic to the propagation of laminar premixed flames, namely the hydrodynamics, Darrieus-Landau (DL), instability, and the diffusional-thermal (DT) instability.

These instabilities are demonstrated in Figure 4.1: in the top row, the flame does not develop wrinkles because the pressure is not high and the deficit reactant, O_2, is also a heavy molecule (Le > 1). In the middle row, the flame exhibits DL instability at high pressure, but not the DT instability (Le > 1). In the bottom row, the flame does not exhibit DL instability (P = 2 atm) but does exhibit DT instability because the deficit reactant, H_2, is a light molecule (Le < 1).

This work was supported by the National Science Foundation.

Keywords:

Cellular instability; Lewis number; Darrieus–Landau; diffusional-thermal

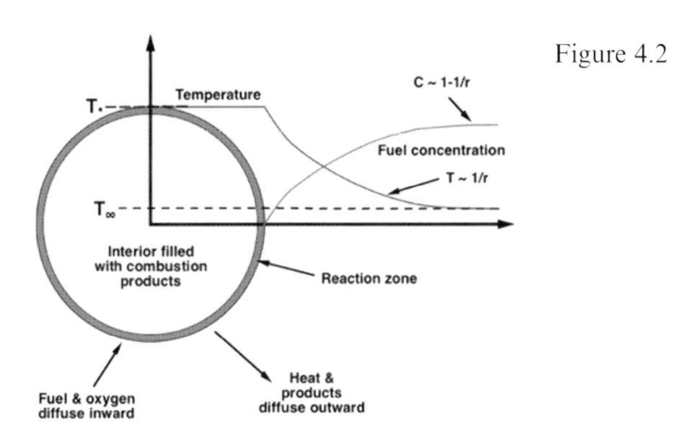

Figure 4.2

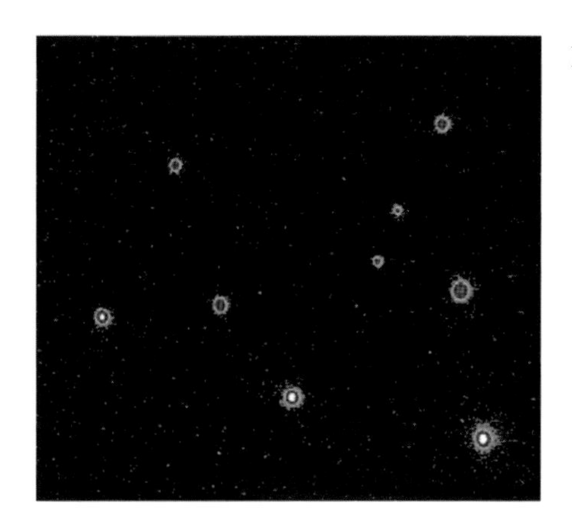

Figure 4.3

4.1.2 Flame Balls

Paul Ronney

University of Southern California

In 1944, Ya. B. Zeldovich predicted the possibility of stationary, steady spherical flames ("flame balls") occurring in premixed gases that are supported exclusively by diffusion of reactants to a reaction zone and diffusion of thermal energy and combustion products away from this reaction zone. Flame balls are valid equilibrium solutions to the governing conservation equations for any combustible mixture, but they were predicted to be unstable (i.e., they would either expand beyond their equilibrium radius or collapse and extinguish) and thus not physically observable.

Forty years later, seemingly stable flame balls were accidentally discovered in drop tower experiments at NASA-Glenn, but the short test times of drop tower experiments and g-jitter effects in parabolic aircraft flight experiments precluded a definite conclusion regarding their stability. This led to the development of the Structure Of Flame Balls At Low Lewis-number (SOFBALL) experiment that flew on Space Shuttle missions STS-83, STS-94, and STS-107. These experiments confirmed the prediction that radiative losses could stabilize flame balls near extinction limits, but it led to a number of surprising observations including mutual repulsion of adjacent flame balls, their extreme sensitivity to tiny gravitational disturbances (on the order of 10^{-6} g), and the uniformity of thermal power among stable flame balls (1–2 watts per ball over a wide variety of mixtures, pressures, and number of balls).

Figure 4.2 is a schematic of a flame ball illustrating representative temperature and concentration profiles.

Figure 4.3 is an image of flame balls in a 7.5% H_2 – 15% O_2 – 77.5% SF_6 mixture taken on the STS-107 mission. Width of field of view is 20 cm. (Flame balls appear to be of different sizes because of [1] differing distances from the camera and [2] differing points in their progress toward extinction at sufficiently small radii.)

This work was supported by NASA.

Keywords:
Microgravity; spherical flames; Lewis number; radiative heat loss

References

1. J. D. Buckmaster, P. D. Ronney, Flame ball drift in the presence of a total diffusive heat flux, *Twenty-Seventh International Symposium on Combustion*, Combustion Institute, Pittsburgh, 1998, 2603–2610.
2. O. C. Kwon, M. Abid, J. B. Liu, P. D. Ronney, P. M. Struk, K. J. Weiland, Structure of Flame Balls at Low Lewis-number (SOFBALL) Experiment, Paper No. 2004-0289, 42nd AIAA Aerospace Sciences Meeting, Reno, NV, January 5–8, 2004.
3. P. D. Ronney, Understanding combustion processes through microgravity research, *Twenty-Seventh International Symposium on Combustion*, Combustion Institute, Pittsburgh, 1998, 2485–2506 (invited paper).
4. P. D. Ronney, M. S. Wu, H. G. Pearlman, K. J. Weiland, Experimental study of flame balls in space: preliminary results from STS-83, *AIAA Journal* 36 (1998), 1361–1368.
5. M.-S. Wu, J. B. Liu, P. D. Ronney, Numerical simulation of diluent effects on flame ball structure and dynamics, *Twenty-Seventh International Symposium on Combustion*, Combustion Institute, Pittsburgh, 1998, 2543–2550.

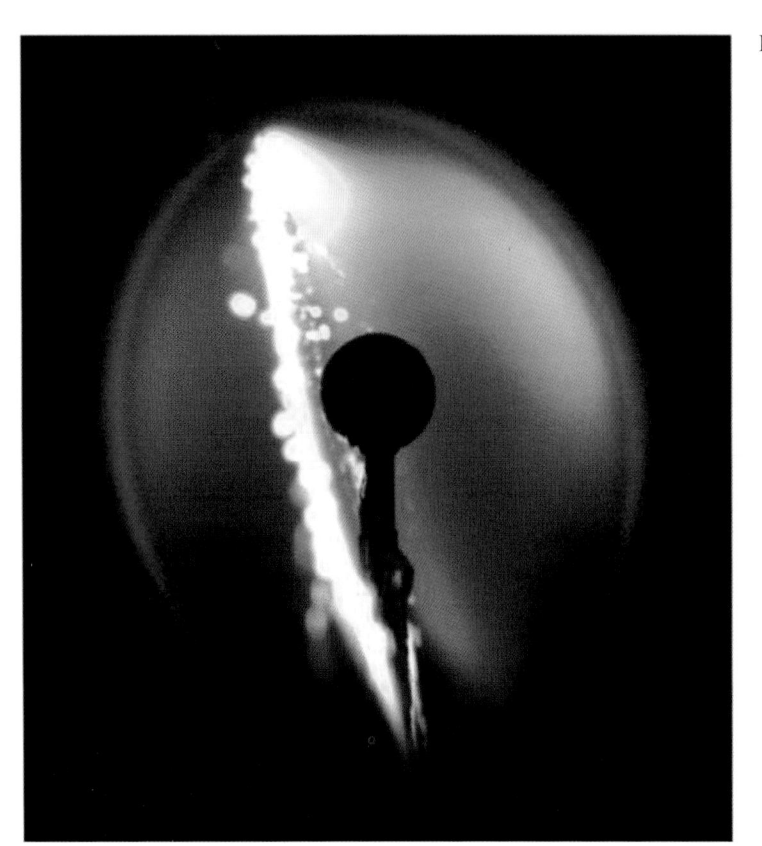

Figure 4.4

4.1.3 Quasi-Steady Microgravity Spherical Ethylene Diffusion Flame

Peter B. Sunderland
University of Maryland

David L. Urban and Dennis P. Stocker
NASA Glenn

Beei-Huan Chao
University of Hawaii

Richard L. Axelbaum
Washington University

This is a color image of an ethylene–air diffusion flame in microgravity. The pressure was 1.01 bar and the ethylene flow rate was 1.51 mg/s. The blue flame sheet and yellow soot are clearly visible. The image was recorded 1.3 s after ignition in the 2.2 s drop tower at NASA Glenn. The camera was a Nikon D100 digital still camera with a resolution of 6 megapixels. The scale is indicated by the 6-mm spherical burner.

The Flame Design International Space Station flight experiment examines the soot inception and extinction limits of spherical microgravity flames. It seeks to improve the understanding of soot inception and control to enable the optimization of oxygen-enhanced combustion and the "design" of non-premixed flames that are both robust and soot free. Tests are conducted with various concentrations of both the injected fuel (i.e., ethylene or methane) and the oxygen-enhanced atmosphere to determine the role of the flame structure on soot inception. The effects of flow direction will be assessed with inverse spherical flames. Flame design explores whether the stoichiometric mixture fraction can characterize soot and flammability limits for non-premixed flames like the equivalence ratio serves as an indicator of those limits for premixed flames.

This work was supported by NASA.

Keywords:
Flame design; radiative extinction; soot

Reference

1. K. J. Santa, B. H. Chao, P. B. Sunderland, D. L. Urban, D. P. Stocker, R. L. Axelbaum, Radiative extinction of gaseous spherical diffusion flames in microgravity, *Combustion and Flame* 151 (2007), 665–675.

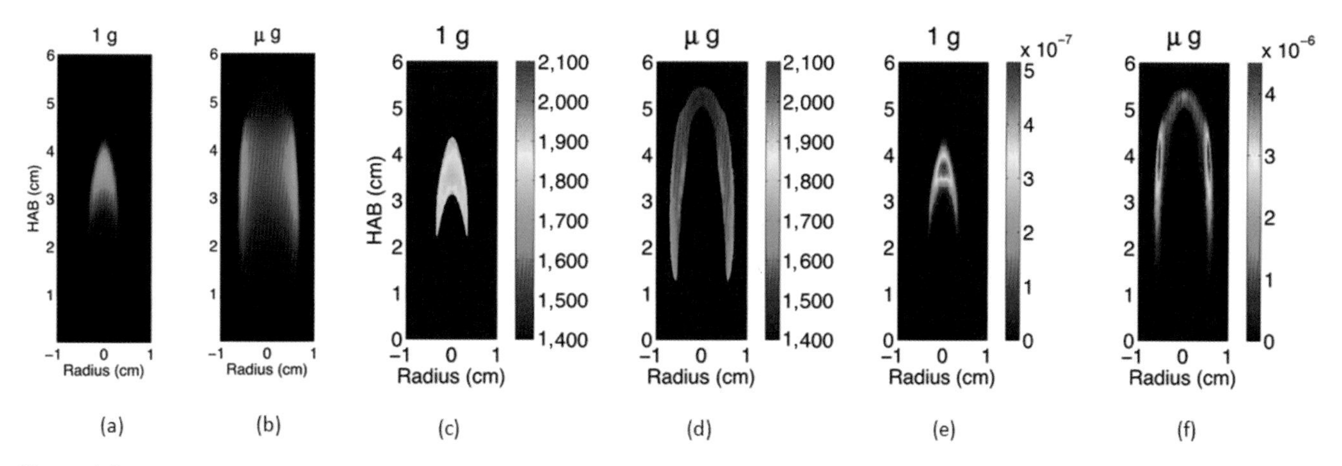

Figure 4.5

4.1.4 Effect of Microgravity on Sooty Co-flow Laminar Diffusion Flames

Davide Giassi, Bin Ma, and Marshall B. Long
Yale University

Dennis P. Stocker
NASA Glenn Research Center

The study of the influence of microgravity on the sooting behavior of axisymmetric co-flow laminar diffusion flames was performed in the Microgravity Science Glovebox, on board the International Space Station during expedition 29/30. Examples of the differences between 1 g and µg flames are illustrated in the photographs above. Figures 4.5a and 4.5b show images of a pure methane flame stabilized in normal and microgravity, respectively. The fuel nozzle has a diameter of 3.23 mm and is surrounded by a 76 mm × 76 mm square duct co-flow. The average flow velocities were 46 cm/s and 18 cm/s for fuel and co-flow, respectively. The images appear green because of the BG-7 color filter that was added to the imaging setup to balance the RGB signal of the color detector.[1]

The cross-sectional soot temperature (in Kelvin), shown in Figures 4.5c and 4.5d, was derived from the collected color images employing an Abel deconvolution and using the color ratio pyrometry technique. Soot volume fraction, shown in Figures 4.5e and 4.5f, was then obtained given the measured temperature and by performing an absolute light intensity calibration.

Due to the lack of buoyancy, in µg a taller and wider flame is produced because the reactant mixing is limited and the inward convection is reduced. The sooty region broadens, following the broadening of the flame, and the peak soot volume fraction increases. The enhanced soot production results in increased thermal radiation losses and hence lower flame temperatures. For these flames, the peak soot temperature in µg is shown to be ~200 K lower than its 1 g counterpart.

Compared to the 1 g flame, the µg counterpart is more diffusion-controlled and has a thicker diffusion layers with more soot production in the wings. From 1 g to µg, the peak of soot volume fraction redistributes from the flame centerline to the wings and the soot growth mode is believed to change from an inception-dominated mode to a surface growth–dominated one.

From accompanying numerical simulations it was seen that, due to the density gradients, the hot flow under normal gravity conditions is accelerated up to 200 cm/s at the far downstream location, while the µg maximum velocity is only 90 cm/s (at the upstream location around the tip of the burner, assuming a parabolic flow profile). The reduction in axial velocity causes longer residence times, allows more time for soot particles to grow, and results in enhanced soot volume fraction.

This work was funded by NASA.

Keywords:
Microgravity; coflow laminar diffusion flame; sooting behavior

Reference

1. B. Ma, S. Cao, D. Giassi, D. P. Stocker, F. Takahashi, B. A. V. Bennett, M. D. Smooke, M. B. Long, An experimental and computational study of soot formation in a coflow jet flame under microgravity and normal gravity, *Proceedings of the Combustion Institute* 35 (2015), 839–846.

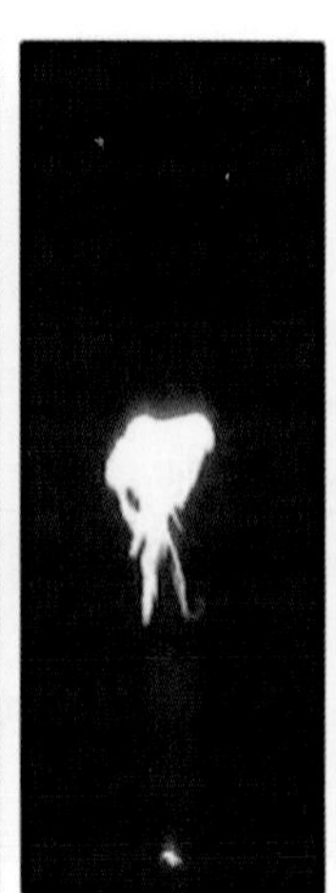

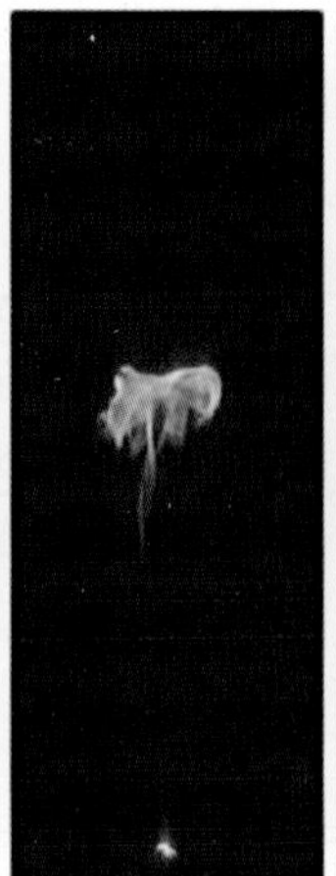

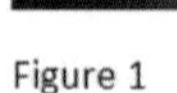

Figure 1 Figure 2

Figure 4.6 Figure 4.7

4.1.5 Steady and Pulsed Sooting Gas Jet Diffusion Flames in Microgravity

James C. Hermanson
University of Washington

Hamid Johari
California State University

Dennis Stocker
NASA Glenn

Uday Hegde
Case Western Reserve University

A fundamental issue in turbulent gas jet diffusion flames is the behavior of large-scale structures that dominate the dynamics of these flames. The research employs a fully modulated fuel injection approach to study both isolated and interacting flame structures. Buoyancy effects are found to be important for these flames, and hence experiments are conducted in microgravity to suppress these effects. The images show ethylene turbulent gas jet diffusion flames in microgravity obtained in the 2.2 s Drop Tower at NASA Glenn. Ethylene is injected at a Reynolds number of 5000 through a 2 cm id nozzle into a chamber filled with a blend of 30% oxygen in nitrogen. Different degrees of interaction between injected flame puffs may be established by varying the injection time for each puff and the time between puff injections.

Figure 4.6 is an image of the sooting, approximately steady-state, diffusion flame. Figure 4.7 shows a sequence of the pulsed case where the fuel is injected for a time of 40 ms. This case studies the dynamics of an isolated flame structure, hence no more fuel is injected until the flame puff is no longer visible. The time between images in Figure 4.7 is 33 ms and the first image is taken approximately 40 ms after injection. The vertical extent of all images is 56 cm. The length of the flame to burnout is related to the ratio of the injection flow rate and radial diffusion rate of oxygen. The reduced burnout length of the pulsed flame indicates better mixing with the surrounding air. The heat release rate and radiative profiles of the flame may be controlled by varying the degree of interaction between the flame structures. The interaction between flame structures also impacts their downstream convection rate.

This research was supported by NASA.

Keywords:
Gas jet; diffusion flames; turbulent; large-scale structures; pulsed; microgravity

References

1. J. Hermanson, H. Johari, D. Stocker, U. Hegde. Buoyancy effects in strongly pulsed turbulent diffusion flames, *Combustion and Flame* 139 (2004), 61–76.
2. J. Hermanson, R. Sangras, H. Johari, J. Usowicz. Effects of coflow on turbulent flame puffs, *AIAA Journal* 40, 7 (2002), 1355–1362.
3. M. Fregeau, J. Hermanson, D. Stocker, U. Hegde. Turbulent structure dynamics of buoyant and non-buoyant pulsed jet diffusion flames, *Combustion Science and Technology*, 182, 3 (2010), 309–330.

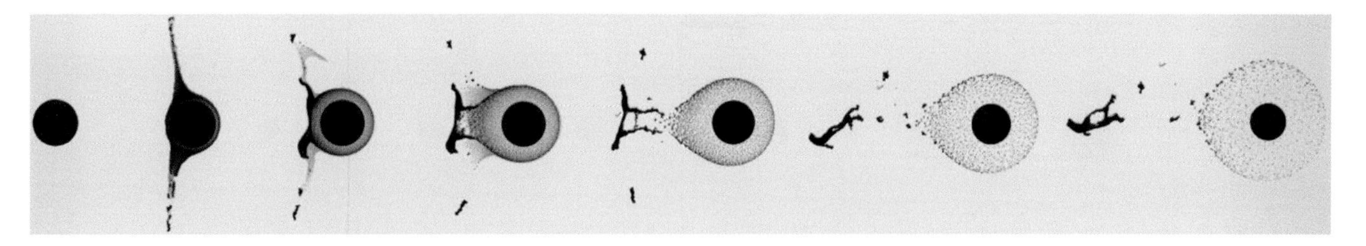

Figure 4.8

4.2 Liquid Fuels

4.2.1 Soot Shell Formation in Microgravity Droplet Combustion

Evan Rose and Vedha Nayagam
Case Western Reserve University

Daniel L. Dietrich
NASA Glenn

The image sequence shows a free-floated decane droplet in microgravity from deployment (first image) to hot flame extinction (last image). Immediately after ignition a dense soot cloud forms close to the droplet (second image). As time proceeds, the soot shell moves slowly outward, as the Stefan flow from the droplet, which pushes the soot away from the droplet, is opposed by the thermophoretic forces on the soot particles, which push the soot toward the droplet (third and fourth images).

As the soot shell expands, the more uniform dense cloud transitions into discrete aggregate particles, and the soot shell expels a very large aggregate particle that has penetrated through the flame due to its increased drag (fifth and sixth images). Once it penetrates the flame, thermophoretic forces and the Stefan flow are both pushing the soot away from the droplet.

Near the radiative extinction of the hot-flame (the last image in the sequence), the symmetric soot cloud is much less dense (the corresponding color of the flame is very dim blue) and located far from the droplet surface.

When the hot flame extinguishes, the soot shell moves rapidly away from the droplet, as the Stefan force is no longer opposed by the thermophoretic force. In this test, a prolonged period of cool-flame burning of the remaining large droplet follows hot-flame extinction.

The experiments reveal a number of interesting aspects of the dynamics of soot particles.[2] These particles normally migrate around the soot-diameter position as they aggregate and eventually move out through the flame. The soot passing through the flame is very evident in the back-lit view because the particle velocity rapidly increases after the particle traverses the reaction zone as a consequence of the suddenly favorable temperature gradient for enhancing the outward thermophoretic velocity, in contrast to the retarding thermophoretic effect for particles inside the flame.

The lower gas density in the vicinity of the flame also helps increase velocities there. Soot particles, thus, are helpful markers of gas motion once thermophoretic effects are subtracted; temperature gradients can be inferred from thermopherotic effects.

Many aspects of soot production and soot-particle histories require further attention. As with many fundamental scientific investigations, this work has uncovered an appreciable number of additional areas worthy of further investigation.

Details of the Flame Extinguishment Experiment are available in Ref. 1.

This research was supported by NASA.

Keywords:
Droplet combustion; soot

References

1. D. L. Dietrich, V. Nayagam, M. C. Hicks, P. V. Ferkul, F. L. Dryer, T. Farouk, et al., Droplet combustion experiments aboard the international space station, *Microgravity Science and Technology* 26 (2014), 65–76.
2. V. Nayagam, J. B. Haggard Jr., R. O. Colantonio, A. J. Marchese, F. L. Dryer, Microgravity n–heptane droplet combustion in oxygen–helium mixtures at atmospheric pressure, *AIAA J* 36, 8 (1998), 1369–1377.

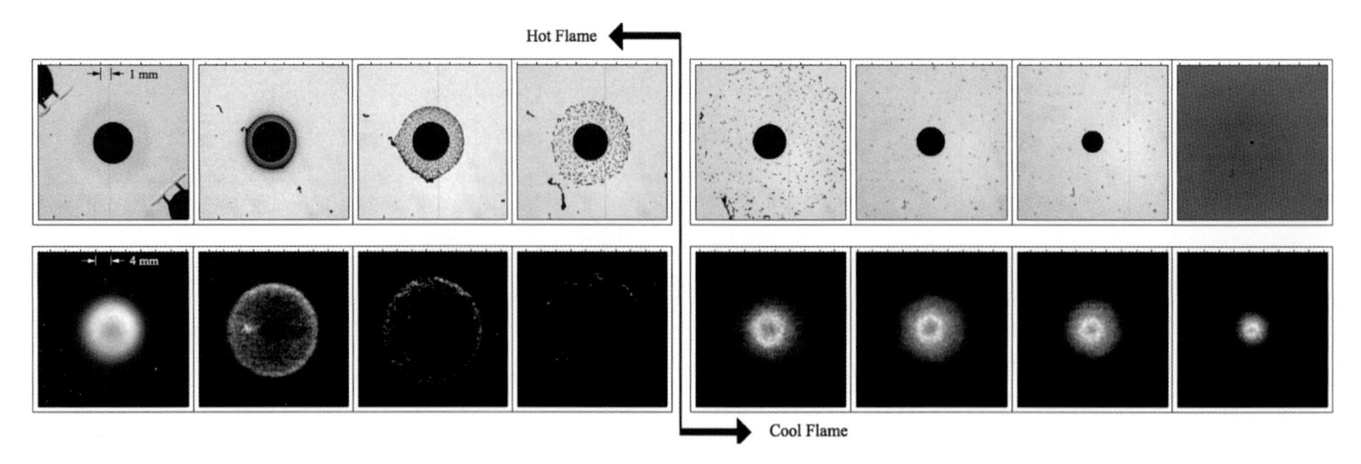

Figure 4.9

4.2.2 Cool-Flame Supported Microgravity Droplet Combustion

Daniel L. Dietrich
NASA Glenn Research Center

Vedha Nayagam
Case Western Reserve University

The image sequence shows a free-floating n-dodecane droplet in microgravity from the moment of ignition to cool flame extinction. For each instant in time there is a backlit image of the droplet and an orthogonal image of the flame. The hot-flame images are the CH* chemiluminescence and the cool flame images are the formaldehyde chemiluminescence.

Immediately after ignition there is a small bright hot flame (the igniters are still visible in the droplet view). Shortly after ignition a dense soot shell forms close to the droplet. The soot particles increase in size and the soot shell grows with time. This corresponds with rapid growth of the hot flame. The flame gets continuously dimmer as it grows and eventually extinguishes due to excessive radiative energy loss (relative to combustion heat release). The hot flame burned for approximately 12 s in the test above, and the images are in 4-s intervals.

Following radiative extinction of the hot flame, a cool flame forms. The cool flame is invisible to the naked eye and required the intensified camera with a formaldehyde filter to visualize the flame. The cool flame temperature is much lower than the hot flame and lies closer to the droplet surface. The soot shell present at hot flame extinction rapidly expands when the hot flame extinguishes, quickly growing in diameter and eventually out of the camera field of view.

The cool flame only burns a fraction of the fuel vaporized from the droplet surface. The rest of the fuel vapor transits the cool flame and can collect far outside the cool flame and recondense to form a fuel cloud that is visible in the backlit view as a darkening of the background and also on a color camera view of the test.

The cool flame burns until the droplet reaches a critical size when it extinguishes leaving a small residual droplet. The cool flame in the image sequence burned for approximately 30 s after the hot flame extinguished, and the images are in approximately 10-s intervals.

Details of the experiment are in Ref. 1, and the cool flame discovery is detailed in Ref. 2.

This research was supported by NASA.

Keywords:
Droplet combustion; cool flames

References

1. D. L. Dietrich, V. Nayagam, M. C. Hicks, P. V. Ferkul, F. L. Dryer, T. Farouk, et al., Droplet combustion experiments aboard the International Space Station, *Microgravity Science and Technology* 26 (2014), 65–76.
2. V. Nayagam, D. L. Dietrich, P. V. Ferkul, M. C. Hicks, F. A. Williams, Can cool flames support quasi-steady droplet burning? *Combustion and Flame* 159 (2012), 3583–3588.

Figure 4.10

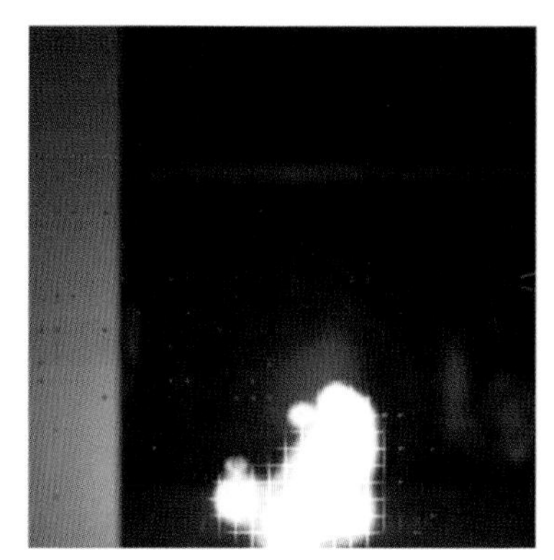

Figure 4.11

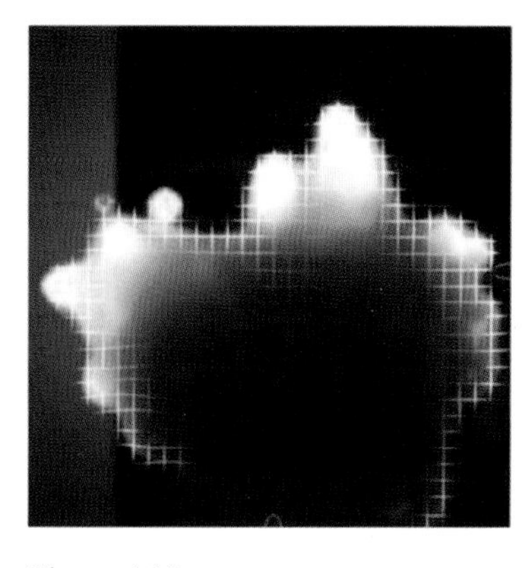

Figure 4.12

Figure 4.13

4.2.3 Flame Spread over a Randomly Distributed Droplet Cloud Aboard Kibo/ISS

Masato Mikami
Yamaguchi University

Masao Kikuchi
JAXA

The left three images show the flame spread behavior over an n-decane 2D droplet array in microgravity aboard the Japanese Experiment Module, Kibo, on the International Space Station (ISS). In this test, 97 droplets were randomly distributed at intersections of a 30 × 30 square lattice with 14 μm silicon carbide fibers placed in a combustion chamber. The spacing between fibers was 4 mm. The initial droplet diameter at ignition was about 1 mm.

One droplet was ignited by a hot-wire igniter to initiate the flame spread at atmospheric pressure (first image). The burning behavior was observed by a digital video camera through the window. The flame spread starts from a small spherical flame around a single fuel droplet and then spreads across the lattice of droplets (second image), until finally a yellow large-scale group flame appears (third image). A blue flame propagating a mixture layer around an unburned droplet is also seen near leading flame regions (top left droplet of third image, for example). The fundamental flame-spread mechanism is similar to that of a linear droplet array in microgravity shown in the small fourth image. However, more dynamic flame-spread behavior appears in randomly distributed droplet clouds. The flame-spread behavior and group-flame formation were investigated for different numbers of droplets and the initial droplet diameters.

Percolation theory has been applied to determine the physics that govern the local flame spread between droplets and the flame spread across the droplet cloud, leading to group combustion.

This research will bridge the gap between the combustion of a small number of droplets and spray combustion.

This research was funded by JAXA under the project entitled "Elucidation of flame-spread and group combustion excitation mechanism of randomly distributed droplet clouds (Group Combustion)."

Keywords:
Flame spread; droplet cloud; group combustion; Kibo/ISS

References

2D Array sequence: www.yamaguchi-u.ac.jp/english/about/news/2017/_6018.html

Percolation model: M. Mikami, H. Saputro, T. Seo, and H. Oyagi, Flame Spread and Group-Combustion Excitation in Randomly Distributed Droplet Clouds with Low-Volatility Fuel near the Excitation Limit: a Percolation Approach Based on Flame-Spread Characteristics in Microgravity, *Microgravity Sci. Technol.*, Vol. 30, No. 4, pp. 419–433 (2018). https://doi.org/10.1007/s12217-018-9603-z

www.combustioninstitute.org/news/advancements-in-combustion/first-combustion-experiments-successfully-started-aboard-the-japanese-experiment-module-kibo/

Linear array image: M. Mikami, H. Oyagi, N. Kojima, Y. Wakashima, M. Kikuchi, S. Yoda, Microgravity experiments on flame spread along fuel-droplet arrays at high temperatures, *Combustion and Flame* 146 (2006), 391–406.

M. Mikami, M. Kikuchi, Y. Kan, T. Seo, H. Nomura, Y. Suganuma, O. Moriue, D. L. Dietrich, Droplet cloud combustion experiment "Group Combustion" in KIBO on ISS, *International Journal of Microgravity Science and Application* 33 (2016), 330208.

Figure 4.14

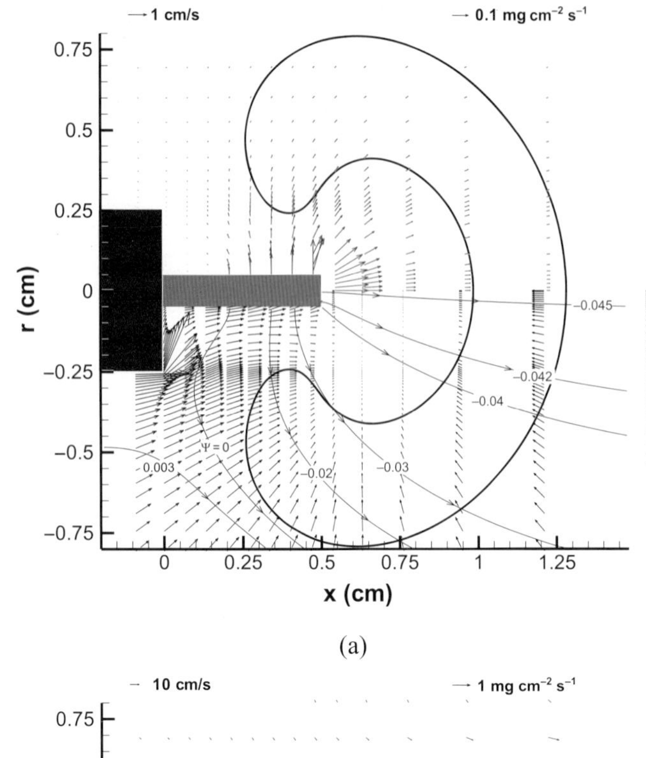

(a)

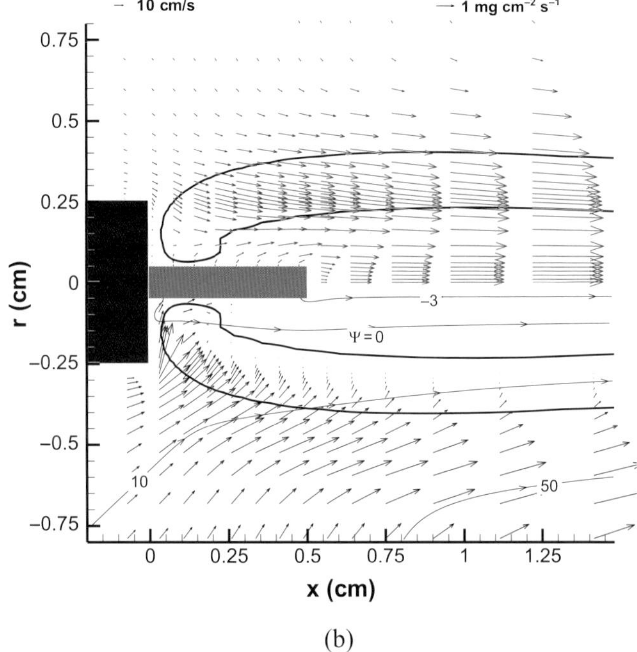

(b)

Figure 4.15

4.2.4 Candle Flames in Normal and Microgravity

James S. T'ien
Case Western Reserve University

Daniel L. Dietrich and Howard Ross
NASA Glenn

Ammar Alsairafi
Kuwait University

Shih-Tuen Lee
National Taiwan University

Figure 4.14 shows that candle flames in normal earth gravity (left) and in microgravity (right) have very different flame shapes, sizes and colors. Figure 4.15 is the numerical simulation of the two cases (with the gravity vector pointed to the left). The visible flames are represented by the fuel reaction rate contour of 5×10^{-5} g/cm^3 s. Upper halves of Figures 4.15(a) and (b) are the velocity vectors, and the lower half shows the streamlines and the oxygen mass flux. The normal-gravity flame is elongated due to buoyancy-induced flow of the combustion products as shown in Figure 4.15(b). Oxygen is entrained into the flame reaction zone primarily by convection and to a lesser extent by diffusion. In the microgravity flame, flow is generated by the Stefan flow resulting from wax evaporation. Comparing Figures 4.15(a) and (b), the magnitude of the flow is much smaller in microgravity (notice the different scales between the two images). In microgravity, Figure 4.15(b) indicates oxygen is supplied to the reaction zone entirely by molecular diffusion. Because these rates are slow, the required fuel supply rate is also small. The flame standoff distance from the wick is therefore large compared with that of the normal gravity case as shown in Figure 4.14. Since the microgravity flame has lower burning and heat release rates, the percentage of radiative heat loss from the flame is larger. The resulting lower flame temperature is believed to be the main reason for a visibly soot-free blue flame shown in Figure 4.14. Details of the experiment are in Ref. 1. Computed candle flames as a function of different gravity levels can be found in Ref. 2. Near-limit candle oscillations were reported in Refs. 1 and 3. Modeling of heat and mass transfer inside the porous wick and wick-trimming effect can be found in Ref. 4.

This research was supported by NASA.

Keywords:
Candle flame; convection; microgravity

References

1. D. L. Dietrich, H. D. Ross, Y. Shu, P. Chang, J. S. T'ien, Candle flames in non-buoyant atmospheres, *Combustion Science and Technology* 156 (2000), 1–24.
2. A. Alsairafi, S.-T. Lee, J. S. T'ien, Modeling gravity effect on diffusion flames stabilized around a cylindrical wick saturated with liquid fuel, *Combustion Science and Technology* 176 (2004), 2165–2191.
3. W. Y. Chan, J. S. T'ien, An experiment on spontaneous flame oscillation prior to extinction, *Combustion Science and Technology* 18 (1978), 139.
4. M. P. Raju, J. S. T'ien, Modeling of candle burning with a self-trimmed wick, *Combustion Theory and Modelling* 12, 2 (2008), 367–388.

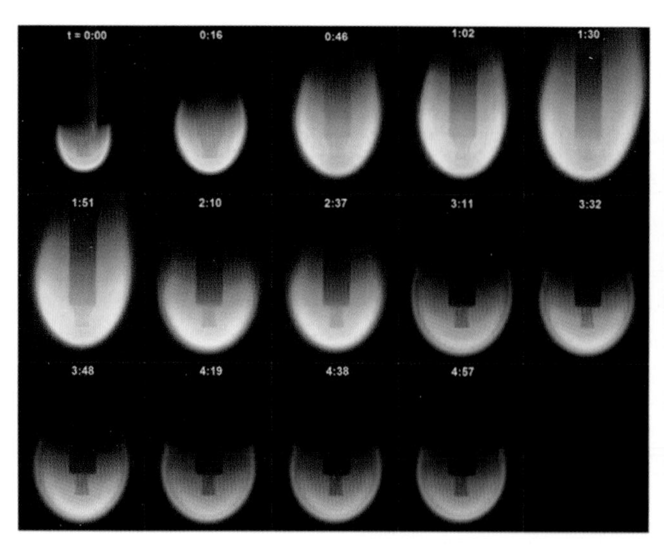

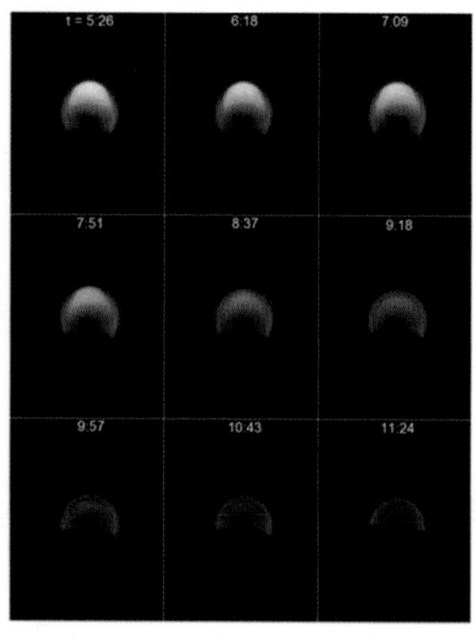

Figure 4.16

4.2.5 Microgravity Candles in the Wind

Paul V. Ferkul

NASA Glenn

Fumiaki Takahashi

Case Western Reserve University

In microgravity, a candle is very sensitive to flow speed and direction.

The candle is comprised of paraffin wax within a 6.5 mm-diameter ceramic tube. A fiberglass string is used as the wick. Flow is generated using a duct having a 7.6 × 7.6 cm cross-sectional area and a volume of 1 liter. Flow direction is from bottom to top in all images and time from ignition (mm:ss) is indicated for each frame.

Left: candlewick points toward the oncoming flow. The initial atmosphere was 17% O_2/N_2 at 1 atm. The flow speed was varied gradually from 5 to 4, 3, 2, 1, and then less than 1 cm/s, at which point the flame goes out as shown in the last image. In this configuration, the flame surrounds the wick and the tip of the candle, promoting fuel vaporization. At the higher flow speeds, the flame is yellow due to significant soot, which is emitted as smoke. At low flow speed, the flame color becomes bluer as the soot diminishes and its shape is more spherical as mass diffusion becomes dominant.

In previous microgravity candle flame studies, a flame could persist even in a quiescent environment. However, the 1-liter flow duct used here was too small to support quiescent combustion. Instead a flow of air always needed to be applied, albeit at a very slow rate (less than 1 cm/s).

Right: candlewick points away from the oncoming flow. The initial atmosphere was 16% O_2/N_2 at 1 atm. The flow in the top row changes from 2 to 1.7 cm/s, the middle row from 1.5 to 1.3 cm/s, and the bottom row from 1.3 to 1.1 cm/s. There is less heat feedback to the candle in this configuration, and so the resulting flame is much smaller compared to the case on the left. The smaller flame burns for a long time because the fuel consumption rate is very low.

The peak heat release rate in the flame stabilization region (base of the flame) is proportional to the flow velocity. Thus, lowering the flow velocity reduces the peak heat release rate and, in turn, heat feedback to the wick, diminishing the fuel vaporization. In other words, the flow velocity and fuel supply rate are coupled in the flame stabilization and extinction processes.

This work was supported by NASA.

Keywords:
Microgravity; candle; flow

References

1. P. V. Ferkul, S. L. Olson, M. C. Johnston, J. S. T'ien, Flammability aspects of fabric in opposed and concurrent air flow in microgravity, 8th U.S. National Combustion Meeting, Paper # 070HE-0218 (2013).
2. F. Takahashi, P. Ferkul, S. Olson, V. R. Katta, Burning characteristics of paraffin and japan wax candle flames in a low-speed oxidizing stream in microgravity, 29th Annual Meeting of the American Society for Gravitational and Space Research, Orlando, FL, November 3–8, 2013.

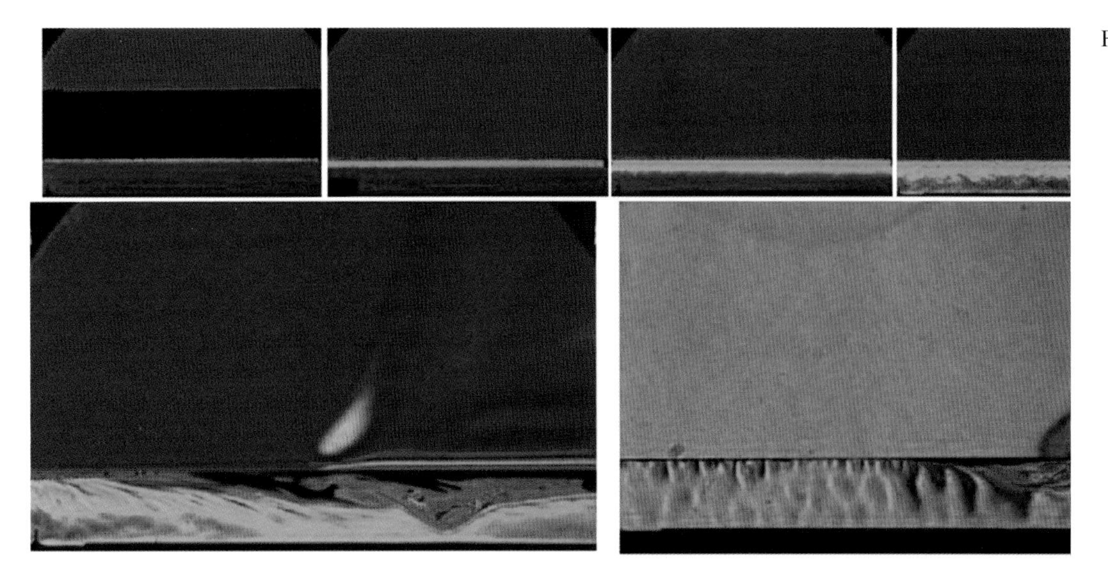

Figure 4.17

4.2.6 Rainbow Schlieren Images of Flame Spread across a Liquid Fuel

Fletcher Miller
San Diego State University

As part of the Spread Across Liquids (SAL) sounding rocket experiment at NASA in the 1990s, rainbow schlieren images were taken of the flame spread process. The flame was ignited at one end of a liquid fuel tray that was 30 cm long × 2 cm wide × 1 cm deep. The fuel was 1-butanol initially at room temperature in these images, and the flame spread is not steady but pulsating at this temperature. Rainbow schlieren filters were used to indicate density (temperature) gradients that arose during the flame spread. All images were taken in normal gravity with a slow opposed air flow.

In the top four images, from left to right, the filter is set to record vertical gradients. The lid on the fuel tray (black) effectively stops evaporation, so that the liquid is all one temperature (blue color). The lid is removed and the surface starts to cool due to evaporation of the 1-butanol. A thermal boundary layer grows downward from the free surface toward the bottom, and by the third image it has grown to approximately 5 mm. In the fourth image, cooling at the surface causes Rayleigh Taylor cells to begin to develop; the cool fluid sinks to the bottom and is replaced by warmer liquid. The entire depth of the liquid is participating in this motion.

In the lower left image, the flame is spreading over the fuel from right to left and the schlieren filter is set to record vertical temperature gradients. Liquid motion is seen several centimeters ahead of the flame front due to surface tension gradients. The flame does not spread steadily but "pulsates": periods of rapid spread are followed by periods of very slow or almost no spread. During each period of slow spread a large vortex builds up beneath the flame driven by surface tension flows on the surface. The vortex nearly reaches the bottom of the pool. When enough fuel vapor has been pulled ahead of the flame by the surface flow, the flame spreads rapidly though this premixed region, and then the process begins again. Due to the much higher index of refraction of the liquid, flow structures are more evident in the liquid phase. However, a thin, high-temperature gradient region is evident right along the liquid surface behind the flame front.

In the lower right image, the filter is set to record horizontal gradients. The gas-phase layer near the surface is no longer evident, since the horizontal gradients in it are very small. However, in the liquid phase, the Rayleigh-Bernard convection cells are now visible as the alternating blue-green strips ahead of the flame. As the flame approaches, those cells are modified by the flow of liquid away from the flame. Once again, the large vortex is seen beneath the flame.

This work was supported by NASA.

Keywords:
Rainbow schlieren; flame spread; liquid pool

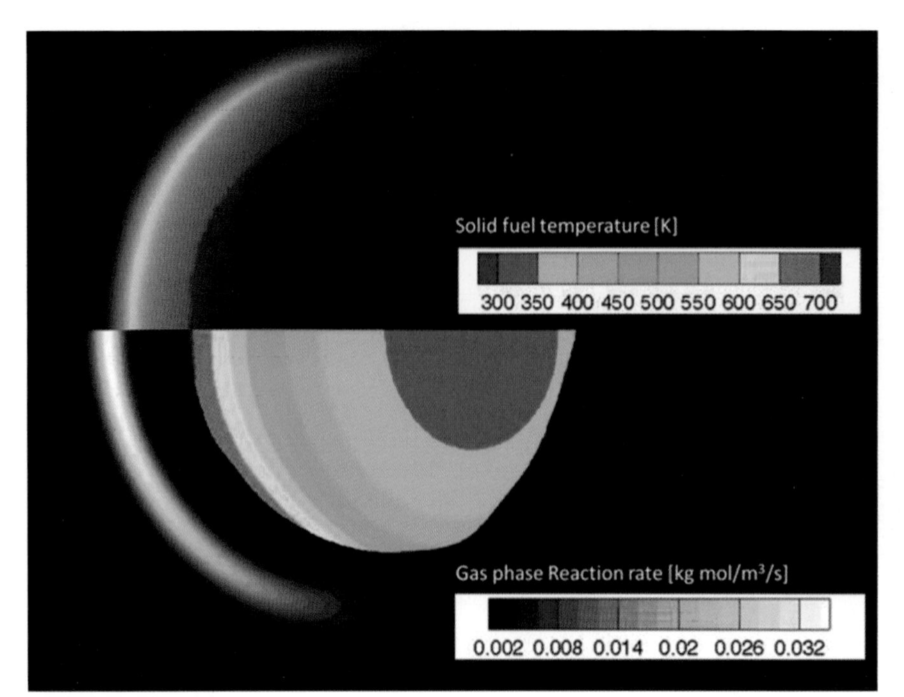

Figure 4.18

4.3 Thick Solid Fuels

4.3.1 Microgravity Sphere Burning with Model Comparison

Makoto Endo, Paul V. Ferkul, and Sandra L. Olson

NASA Glenn

James S. T'ien

Case Western Reserve University

Still image of burning Poly (methyl methacrylate) in microgravity condition aboard International Space Station (top, obtained through the Burning And Suppression of Solid fuels (BASS) project), compared with numerical model [2] that incorporates the effect of surface regression and solid surface heat balance (bottom). Forced flow velocity is 12 cm/s (left to right), and the oxidizer consists of an oxygen/nitrogen mixture (17%/83% by mole, respectively).

The top half of the image shows the silhouetted shape of the regressing sphere and the blue flame stabilized around the forward half of the sphere. The bottom half of the image shows the numerical model predictions. The gas-phase blue color corresponds to the gas-phase reaction rate. The flame standoff distance is overpredicted by the model. This may be due to a mismatch in the timing of the solid-phase heating between experiment and model. The rainbow colors represent the solid fuel temperature distribution, showing how the flame is heating the sphere.

This work was supported by NASA.

Keywords:
PMMA sphere; microgravity; flame growth; numerical model

References

1. S. L. Olson, P. V. Ferkul, S. Bhattacharjee, F. J. Miller, C. Fernandez-Pello, S. Link, et al., Results from on-board CSA-CP and CDM sensor readings during the Burning and Suppression of Solids – II (BASS–II) experiment in the Microgravity Science Glovebox (MSG), ICES–2015–196, 45th International Conference on Environmental Systems, July 12–16, 2015, Bellevue, WA.
2. M. Endo, Numerical modeling of flame spread over spherical solid fuel under low speed flow in microgravity: model development and comparison to space flight experiments. Electronic Dissertation. Case Western Reserve University, 2016. https://etd.ohiolink.edu/

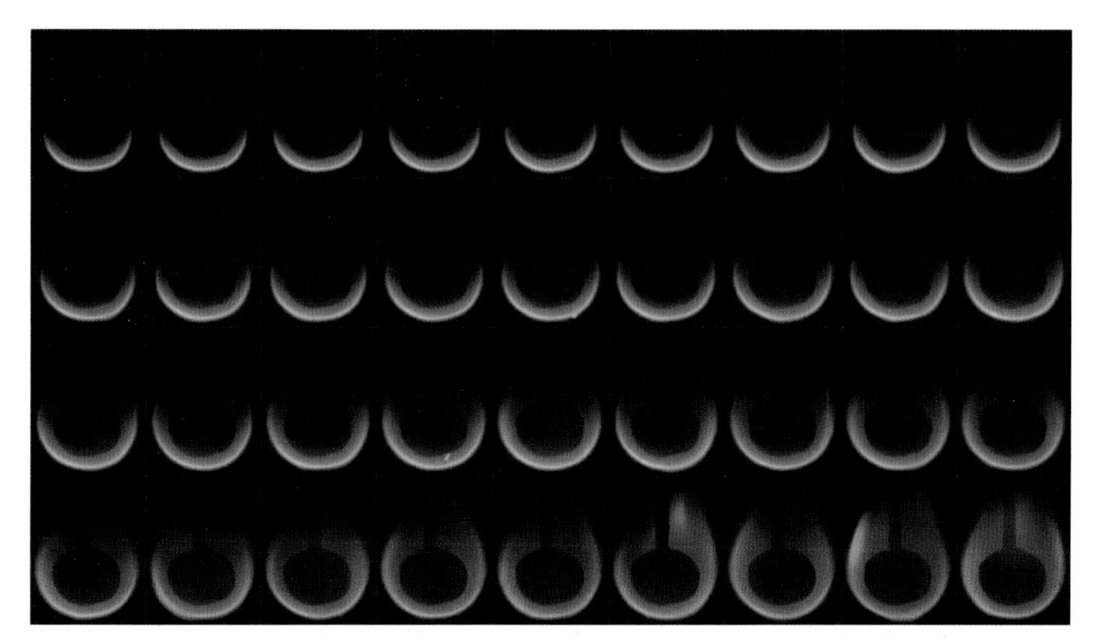

Figure 4.19

4.3.2 Burning Sphere in Microgravity

Paul V. Ferkul, Makoto Endo, and Sandra L. Olson

NASA Glenn

James S. T'ien and Michael C. Johnston

Case Western Reserve University

Still images of burning Poly (methyl methacrylate) in microgravity condition aboard International Space Station obtained through the Burning And Suppression of Solid fuels (BASS) project. Forced flow velocity is 12 cm/s (upward), and the oxidizer consists of an oxygen/nitrogen mixture (17%/83% by mole, respectively). The initial diameter of the solid fuel was 2 cm, and each image is about 1.5 s apart.

A cast PMMA sphere of initial diameter of 2 cm was supported by a rod inserted through the back side. The sample is placed in the flow tunnel. A heated wire igniter is placed at the forward stagnation region (i.e., upstream). Once the flame is established, the igniter is retracted. As can be seen in the top row of images, the initial flame is localized around the forward half of the sphere and is all blue.

As time goes on, the flame tip spreads downstream to the sides of the sample. The flame develops an inner soot region (orange-yellow area) by the third row of images. This indicates that the flame is getting stronger as the conductive heat flux to the solid phase decreases with time as the sample interior is heated up.

By the forth row, the flame has enveloped the sphere. The flame is stable and smooth for this low Reynolds number flow, except for intermittent vapor jetting from small bubbles rupturing at the fuel surface as it burns.

Such series of high-resolution images are utilized for model development by providing validation data of flame growth rate, flame-to-fuel distance, and the solid surface regression rate.

This work was supported by NASA.

Keywords:
PMMA sphere; microgravity; flame growth

References

1. M. Endo, Numerical modeling of flame spread over spherical solid fuel under low speed flow in microgravity: model development and comparison to space flight experiments, PhD Dissertation, Case Western Reserve University, February 2016.
2. M. Endo, J. S. T'ien, P. V. Ferkul, S. L. Olson, Experimental data analysis and numerical modeling of flame spread on a PMMA sphere in microgravity, presented at the 32nd annual meeting of the American Society for Gravitational and Space Research (ASGSR), Cleveland, OH, October 26–29, 2016.
3. S. L. Olson, P. V. Ferkul, S. Bhattacharjee, F. J. Miller, C. Fernandez-Pello, S. Link, et al., Results from on-board CSA-CP and CDM sensor readings during the Burning and Suppression of Solids – II (BASS–II) experiment in the Microgravity Science Glovebox (MSG), ICES-2015-196, 45th International Conference on Environmental Systems, July 12–16, 2015, Bellevue, WA.

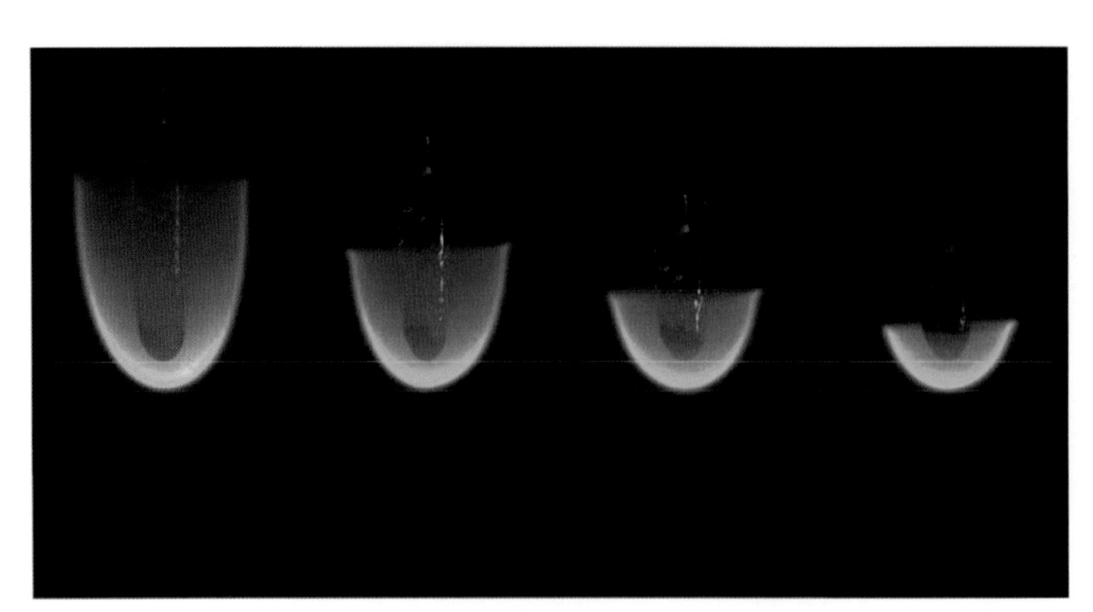

Figure 4.20

4.3.3 Microgravity Flames Stabilized around the Stagnation Tip of a Clear Polymethylmethacrylate (PMMA) Rod

Sandra L. Olson and Paul V. Ferkul

NASA Glenn Research Center

During a test in the BASS-II hardware aboard the International Space Station, a 0.95 cm-diameter PMMA rod is burned in a very slow concurrent flow.[1] Flow direction is up in each of the four images. From left to right, the effect of decreasing flow speed is shown: 2 cm/s, 1.5 cm/s, 1.1 cm/s, and 1 cm/s. Allowing the rod sufficient time to reach a stable solid-phase thermal profile after each flow step, the flame became exceptionally stable, and the image was taken. The last flame was held at 1 cm/s flow for 100 s until extinction finally occurred because of oxygen depletion. The initial oxygen concentration in the chamber was 14.8% O_2 at 1 atm., and the final oxygen concentration was 13.6% O_2. These oxygen levels are well below the minimum for which this fuel will burn on Earth (15% O_2).

The rod tip is aligned in all four images as shown by the fine yellow line. Despite the flow changes, the stagnation flame standoff distance remained the same, as shown by the fine orange line. Once the solid phase reached a steady-state thermal profile, the rod regression rate became constant. The primary effect of decreasing flow is a proportional decrease in the flame length. As the quantity of available oxygen was reduced (via reduced flow and slowly dropping ambient oxygen concentration), the size of the flame sheet shrunk to match.

Also visible in each image is the blackened tip of the rod that has an elliptical shape. The clear portion of the rod shows a dense bubble layer that spans the extent of the flame. Near the quenching extinction (right-most image), the bubble layer length had shrunk to approximately the same size as the rod diameter. This is an indication of the hemispherical shape of the flame, for which the very low convective flow speeds and mass diffusion speeds are approximately equal.

Prior to extinction, the flame length oscillated at a frequency of 1.1 Hz. These oscillations are very similar to those observed for candle flames in a quiescent environment,[2] since the 1 cm/s convective flow is the same level as mass diffusion speeds in the vicinity of the candle flame.

This work was supported by NASA.

Keywords:
Microgravity; concurrent flow; PMMA; axial rod; stagnation flow

References

1. S. L. Olson and P. V. Ferkul, Microgravity flammability boundary for PMMA rods in axial stagnation flow: experimental results and energy balance analyses, *Combustion and Flame* 180 (2017), 217–229.
2. D. L. Dietrich, H. D. Ross, Y. Shu, J. S. T'ien, Candle flames in non-buoyant atmospheres, *Combustion Science and Technology* 156, 1 (2000), 1–24.

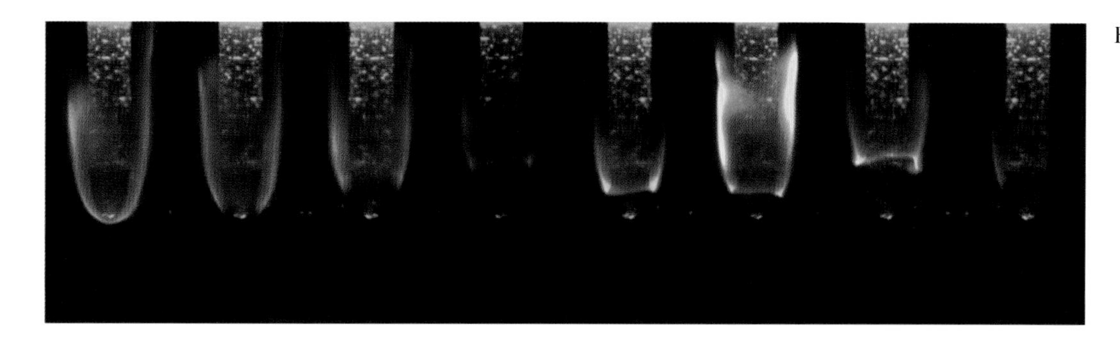

Figure 4.21

4.3.4 Microgravity Concurrent Flame Blowoff from the Tip of a PMMA Rod

Sandra L. Olson
NASA Glenn

This sequence of images is taken from a 5.18 s drop tower test burning a 1.27 cm-diameter PMMA rod in 15% oxygen at 30 cm/s concurrent flow (up in the images) at 1 atm. pressure. The eight frames are 1/30 s apart. The rod interior is illuminated with a green LED to show the bubble layer that develops beneath the flame as the material degrades in depth. While the end of the rod is blackened by soot, some green illumination can be seen right at the tip.

Prior to the drop, the rod was ignited with a retractable hot wire coil at the rod tip in an 18% oxygen atmosphere. Once ignited, the flame was allowed to grow and heat the rod. The drop was triggered and the gas flow was switched to 15% oxygen.

Blowoff occurs shortly after the drop starts, and is triggered by the opening of a small hole that forms at the stagnation tip of the rod, as shown in the second image. This hole is caused by the Damkohler number reaching a critical value where the reactions cannot occur fast enough in the finite thickness of the reaction zone for the given flow time. This hole quickly grows to the scale of the rod diameter as the stagnation region of the flame destabilizes. Due to the destabilization, the flame quickly blows downstream.

The flame flashes back forward a few millimeters and develops a bright blue flame base that anchors briefly to the sides of the cylinder in the slower-flow environment in the rod boundary layer (fifth image). This bright blue base is interpreted as the outer lip of the triple flame at the flame's leading edge of an opposed flow flame. The downstream section of the flame becomes sooty (sixth image), as the newly anchored flame burns out the unburned fuel that has leaked through the open tip of the flame. However, the flame base cannot sustain the side-stabilization and blows downstream, extinguishing the flame.

Some blowoff tests show a prolonged oscillation of the anchoring flame ring prior to blowoff. The triple flame base resides within the boundary layer along the side of the rod. The still hot rod tip continues to produce fuel vapor, and a flammable mixture is formed within the boundary layer. The bright blue triple flame moves forward through this layer, consuming the mixture. Once the mixture is consumed, the flame is blown downstream to a lower flow deeper in the boundary layer where it can stabilize. The cycle repeats a number of times, but each time the flame is blown slightly further downstream as the mixture becomes leaner.

The flame speed during the oscillations is on the order of gas-phase premixed flame speeds. Interestingly, both the maximum flame speed and the brightest blue triple flame region occur during the last few oscillations. The blue chemilumienscene is an indication of the reaction rate, and the flame speed is fastest near stoichiometric mixtures. So, the fact that the flame speed increases as the rod cools indicates that the flammable mixture layer was initially fuel rich but becomes progressively leaner as the rod cools. Once the blown-off rod tip cools sufficiently, the flame can no longer find a thick enough flammable mixture layer forming upstream, and the flame extinguishes.

This work was supported by NASA.

Keywords:
PMMA; rod; axial flow; stagnation point; blowoff; microgravity

References

1. J. W. Marcum, S. L. Olson, P. V. Ferkul, Mixed convection blowoff limits as a function of oxygen concentration and upward forced stretch rate for burning PMMA rods of various sizes, 47th International Conference on Environmental Systems, Charleston, SC, July 16–20, 2017.
2. S. Olson, P. Ferkul, Microgravity flammability boundary for PMMA rods in axial stagnation flow: experimental results and energy balance analyses, *Combustion and Flame* 180 (2017), 217–229.
3. S. L. Olson, P. V. Ferkul, J. W. Marcum, Analysis of high speed video of PMMA rod blowoff, 33rd annual meeting of the American Society for Gravitational and Space Research (ASGSR), Renton, WA, October 25–28, 2017.

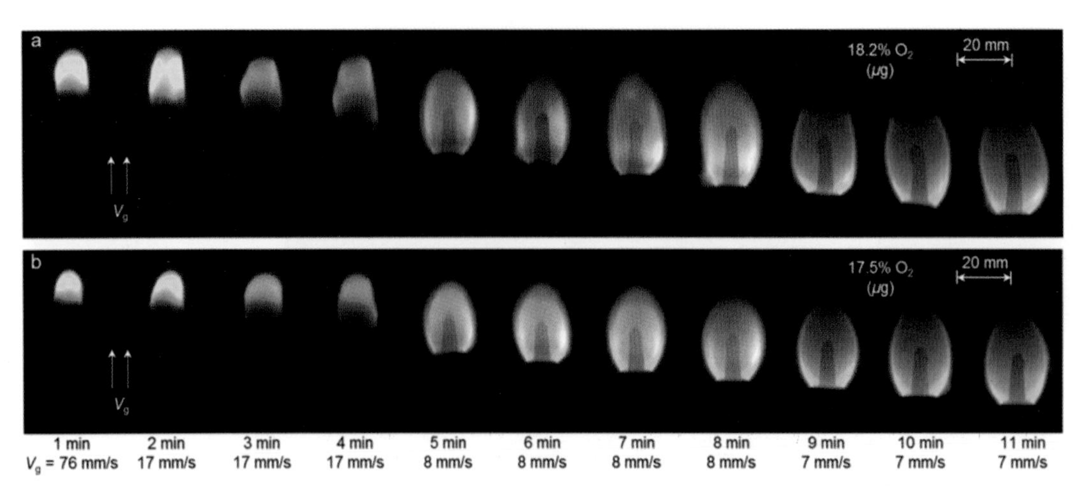

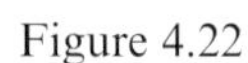

Figure 4.22

4.3.5 Microgravity Flame Spread over a Black Polymethylmethacrylate (PMMA) Rod

Xinyan Huang and Carlos Fernandez-Pello

University of California, Berkeley

Sandra Olson and Paul Ferkul

NASA Glenn

Fire safety in microgravity has always been a concern in space travel. Very few flame-spread experiments, particularly on thick fuels, have been conducted in spacecraft environments because of the limited long-term microgravity facilities. The presented fire image shows part of the microgravity flame-spread experiments, Burning And Suppression of Solids - II (BASS-II), conducted on the International Space Station.

The flame spread over the 6.4 mm-thick black PMMA rod was tested under low-velocity opposed flows (V_g).[1] These fire images show the opposed flame spread under a gas flow of low oxygen concentration by volume of 18.5% (upper row) and 17.2% (lower row). The interval between these images is 1 minute. As the opposed flow is slowly decreased from 76 mm/s to 7 mm/s, the flame changes from yellow to blue and becomes wider. The measured flame spread rate first increases when the opposed flow velocity is reduced from 76 mm/s to 17 mm/s, and then it decreases as the opposed flow velocity is further reduced. Eventually, flame extinction occurs when the flow velocity is reduced to 6 mm/s.

The observed correlation between flame spread rate and the opposed flow velocity has only been predicted theoretically[2] but not been observed experimentally for thick fuels until this work. In normal gravity, the strong upward buoyant flow (about 300 mm/s) generated by the flame dominates over the low-velocity forced flow. As a consequence, the flame-spread rate measured on Earth is almost independent of the flow velocity,[3] which prevents the study of the flammability of materials in spacecraft environments.

This group of photos also show that in microgravity flame spread could be sustained at oxygen concentration of 17.5% or lower. However, in normal gravity, the minimum oxygen concentration to sustain a flame in the same PMMA sample is 18.3%. These novel results suggest that under certain environmental conditions there could be a higher fire risk and a more difficult fire suppression in microgravity than on Earth, which would have significant implications for spacecraft fire safety.

This work was supported by NASA.

Keywords:
Low flow velocity; opposed flow; oxygen concentration

References

1. S. Link, X. Huang, C. Fernandez-Pello, S. Olson, P. Ferkul, The effect of gravity on flame spread over PMMA cylinders, Scientific Reports, Vol. 8, Article 120 (2018).
2. C. Di Blasi, S. Crescitelli, G. Russo, C. Fernandez-Pello, Predictions of the dependence on the opposed flow characteristics of the flame spread rate over thick solid fuel, *Fire Safety Science* 2 (1989), 119–128.
3. C. Fernandez-Pello, S. Ray, I. Glassman, Downward flame spread in an opposed forced flow, *Combustion Science and Technology* 19 (1978), 19–30.

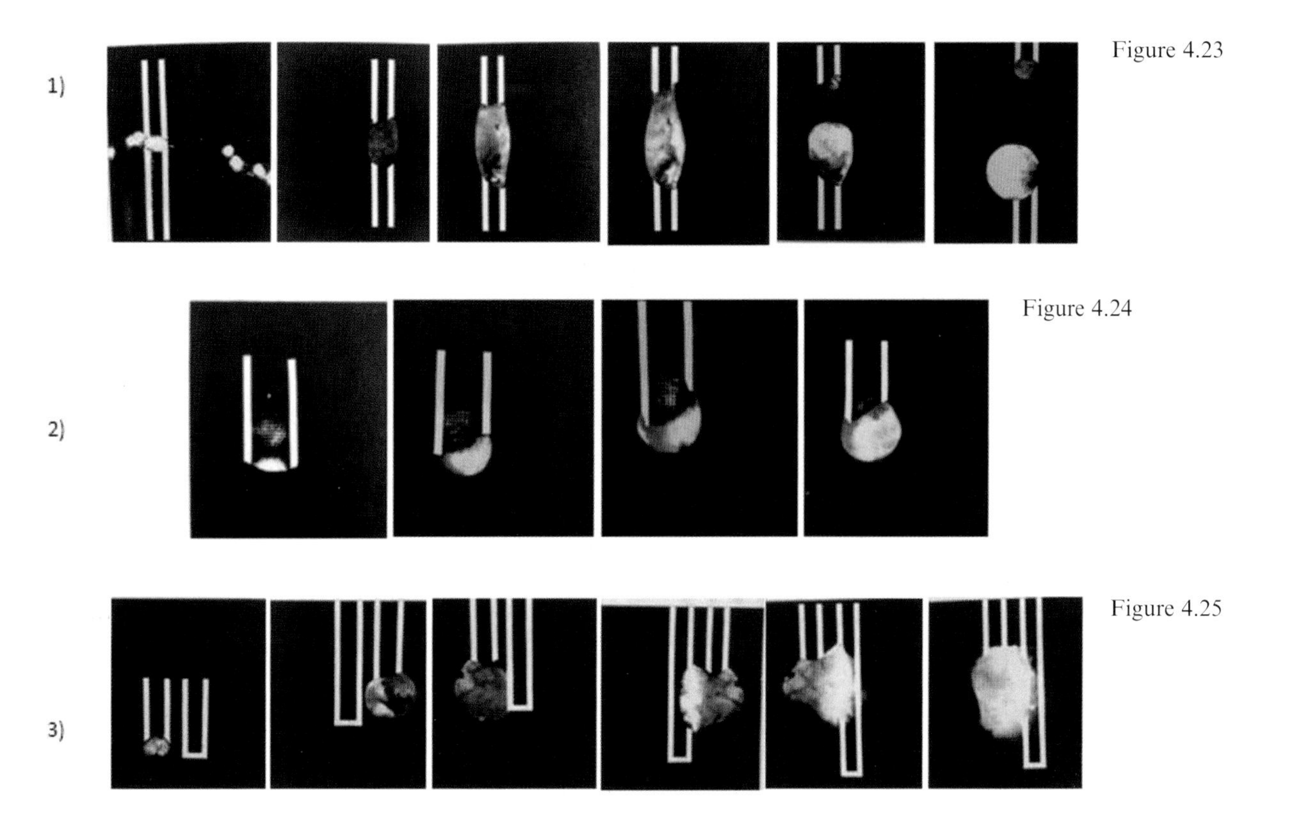

Figure 4.23

Figure 4.24

Figure 4.25

4.3.6 Heterogeneous Burning of Bulk Metallic Materials in Pure Oxygen in Reduced Gravity

Ted Steinberg

Queensland University of Technology

Figure 4.23 shows center ignition/burning of a 0.32 cm-diameter cylindrical iron rod in microgravity (6.9 MPa white lines added to show sample boundaries).[1,2] The igniter was Al-Pd wire, and burning progresses up the test sample until molten when surface tension pinches the liquid column to create two spherical burning regions that continue to consume the sample (over 1.2 s). Figure 4.24 shows a "rolled-up" 316-Stainless Steel mesh burning in microgravity over 0.08 s (1.7 MPa).[1,2] The burning liquid mass rises into the mesh by capillary action before burning it from the inside out. The complete sample was consumed, and after the test, unlike in a normal gravity test, green oxide is distributed all around due to the higher temperatures in the reaction zone (leading to volatilization of the chromium in the stainless steel). Figure 4.25 shows the ignition (3.5 MPa over 1.5 s) of a 0.32 cm-diameter cylindrical solid iron rod in microgravity that produces the typical (in microgravity) spherical burning molten mass that then contacts, melts, and ignites a 0.32 cm-diameter cylindrical 2219 Aluminum rod, clearly showing the different burning characteristics of these two metals.[1,2]

These experiments studied flammability of metallic materials in relation to combustion mechanisms and fire safety and established the heterogeneous burning mechanism for most metals. The work also demonstrated that bulk metallic materials are consumed much faster in microgravity than in normal gravity – the first time an increased burn rate was shown for a material in microgravity.

This work was supported by NASA.

Keywords:
Metals combustion; burning metals; fire safety; heterogeneous combustion; microgravity

References

1. T. A. Steinberg, D. B. Wilson, F. J. Benz, The burning of metals and alloys in microgravity, *Combustion and Flame* 88 (1992), 309–320.
2. T. A. Steinberg, D. B. Wilson, F. J. Benz, Microgravity and normal gravity combustion of metals and alloys in high pressure oxygen, Flammability and Sensitivity of Materials in Oxygen-Enriched Atmospheres: Vol. 6, ASTM STP1197, 1993.

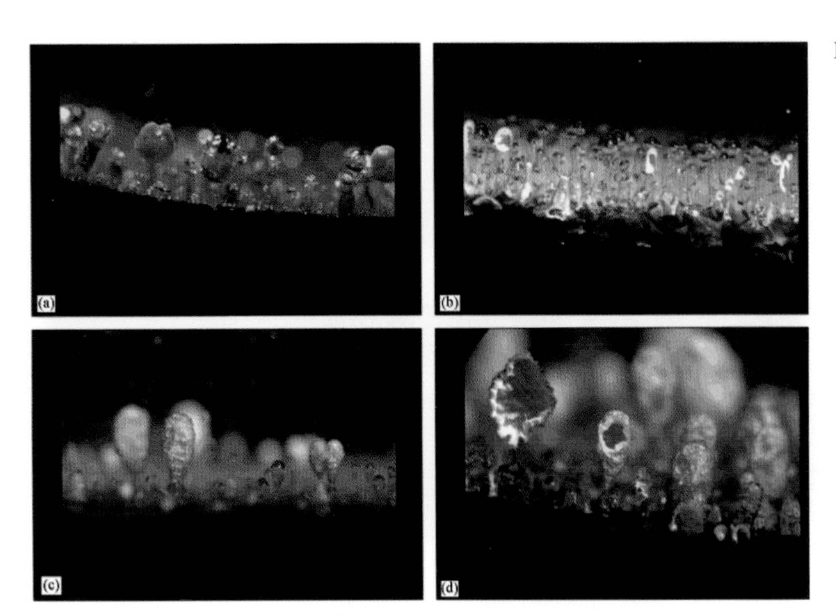

Figure 4.26

4.3.7 PMMA In-Depth Bubble Formation

Sandra L. Olson
NASA Glenn

James S. T'ien
Case Western Reserve University

The characteristics of molten layers of burning materials are not well understood. The samples tested were 2.4 cm-thick PMMA slabs (30 cm × 30 cm) heat-formed into large radius of curvature samples. These large-radius-of-curvature samples had very slow regression rates (low Peclet numbers). An extensive layer of polymer material above the glass transition temperature (105°C) was observed during the burning of these samples in a ceiling fire configuration. A bubble layer is formed beneath the burning surface. The figure shows magnified images of the resolidified bubble layers: cross-section cuts through the sample to show the in-depth bubble formation for four different sample radii: (a) 5 cm, (b) 20 cm, (c) 50 cm, and (d) 100 cm.

The depth of the bubble layer increases as the radius of curvature increases and the Peclet number decreases. The bubbles become less glassy and more opaque. The opaque, wrinkled bubbles seen in the cross-section images are the MMA vapor bubbles after the samples resolidify after the test. The opaque wrinkling of the bubbles is believed to be due to local fracturing of the cooling/solidifying molten polymer–vapor interface and/or crystal growth of crystalline PMMA as the pressure within the vapor bubble decreases due to cooling and condensation of the MMA vapor. This give the cut bubbles a geode appearance.

Unique phenomena associated with this extensive glass layer included an initial substantial swelling of the burning surface (4 mm) due to the initial formation of the bubble layer and migration and/or elongation of the bubbles normal to the hot surface through the viscous molten polymer. The bubble layer acted to insulate the polymer surface by reducing the effective thermal conductivity of the solid.

As the material thermally degrades in depth at temperatures above 225°C, the random scission of the polymer chains releases gaseous MMA monomer (B.P. 101°C, just below the glass transition temperature) that forms bubbles in the molten polymer. These bubbles are at greater than ambient pressure. As bubbles reach the burning surface (~360°C), these pressurized bubbles rupture and a jet of MMA fuel vapors is released. The fuel ignites as it passes through the flame sheet, causing a flame jet and a characteristic sizzling audible signature. The vapor jets add to the effective fuel vaporization rate and enhance burning in slow convective flow environments by entraining ambient air flow.

This work was supported by NASA.

Keywords:
PMMA; bubble layer; Peclet; glass transition

Reference

1. S. L. Olson, J. S. Tien. Near surface vapor bubble layers in buoyant low stretch burning of polymethyl-methacrylate, *Fire and Materials* 23 (1999), 227–237.

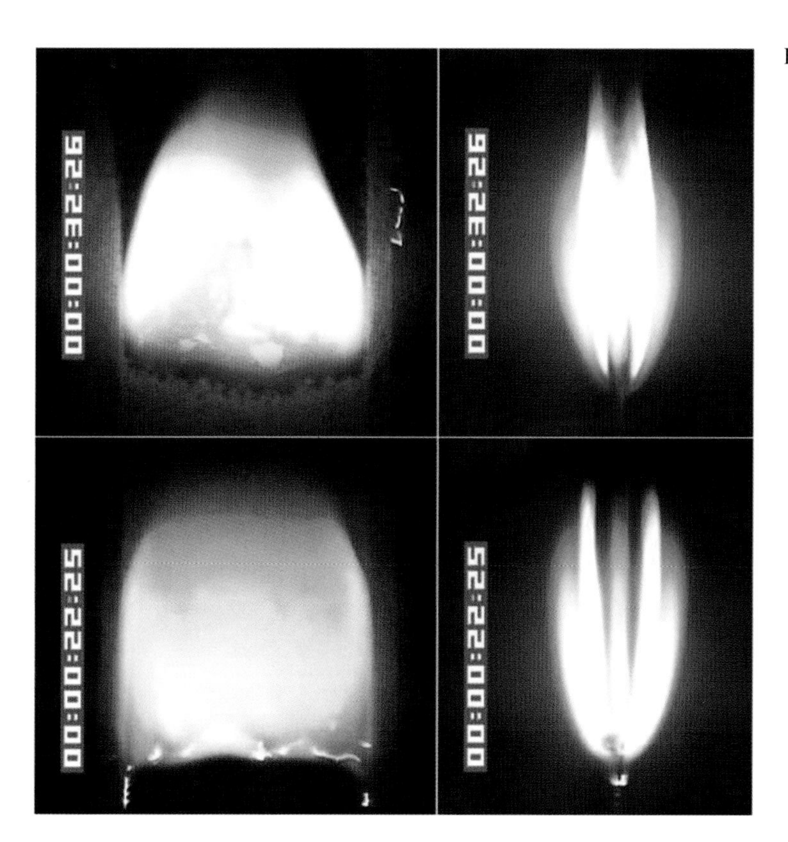

Figure 4.27

4.4 Thin Solid Fuels

4.4.1 Experimental Comparison of Opposed and Concurrent Flame Spread in a Forced Convective Microgravity Environment

Sandra L. Olson
NASA Glenn

Fletcher J. Miller
San Diego State University

A series of 5.18 s drop tower tests were performed using a very thin cellulose samples with an igniter wire on either the upstream or downstream end of the 5 cm-wide by 10 cm-long sample. Top view (left) and side view (right) images are shown for two tests at 34% O_2, 10.2 psia, and 30 cm/s. The spread rates are nearly identical, but the flame structure is quite different for the opposed flow flame (top images) and the concurrent flow flame (bottom images).

The opposed-flow flame has a fairly flat dark pyrolysis front ahead of the sooty region of the flame. The material burns out quickly, with only some lacy char within the luminous flame. The edge view shows the classic blue leading edge, followed by a long luminous merged inner tail that fills the open space behind the burnt-out fuel. An outer magenta halo flame curves inward and is shorter than the sooty tail. For opposed flow, the flame leading edge reaches a steady spread rate almost immediately (<1 s) while the flame size continues to grow for a few seconds.

For the concurrent flame, the flame base (burnout) is quite flat with a long pyrolysis region beneath the flame. The concurrent flame exhibits a separated inner sooty tail structure, and the flame standoff distance is significant. There is again an outer magenta halo flame. In the concurrent case, the flame base remains stationary for ~1 s after ignition as the flame length develops. The flame base then accelerates to a steady spread rate within ~1 s once it starts to propagate.

This work was funded by NASA.

Keywords:
Opposed flow; concurrent flow; flame spread; microgravity; cellulose

Reference

1. S. L. Olson, F. J. Miller, Experimental comparison of opposed and concurrent flame spread in a forced convective microgravity environment, *Proceedings of the Combustion Institute* 32, 2(2009), 2445–2452.

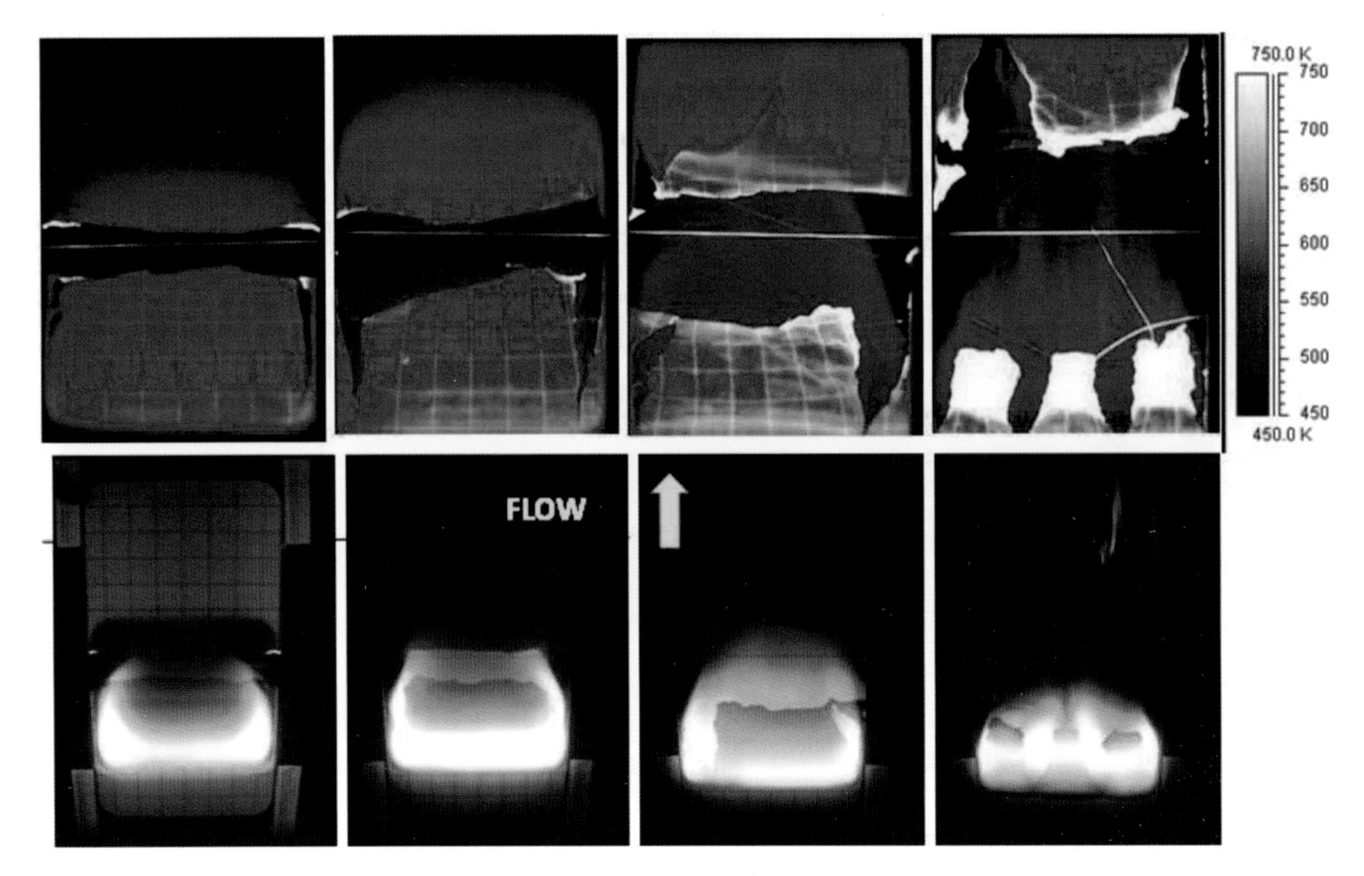

Figure 4.28

4.4.2 Visible and Infrared Images of Flow Effects on Simultaneous Upstream and Downstream Flame Spread in Microgravity

Sandra L. Olson
NASA Glenn

Osamu Fujita
Hokkaido University

Masao Kikuchi
JAXA

Takashi Kashiwagi
NIST

The effect of low-velocity forced flow on microgravity flame spread is shown in these images: (top) quantitative infrared images of the burning surface; and (bottom) color images of the surface view. The IR image has a shorter field of view and is the mirror image of the visible view. The images are roughly to scale, with a 1-cm grid printed on the paper. The scale is visible in the IR, since the emissivity is slightly higher on the grid lines. A color temperature scale for the IR images is shown on the right. The forced flow is up in these images.

These images were taken at the end of the JAMIC 10 s drop tower in Hokkaido, Japan. In these tests, the flame is able to spread both upstream and downstream from a central ignition on the fuel surface. The Whatman 44 ashless filter paper cellulose fuel samples were ignited in microgravity using a straight hot wire across the center of the 75 mm-wide by 140 mm-long samples. Red LEDs illuminate the visible sample. Very thin thermocouple wires are also present, and are visible in some IR images. The flame spread is reasonably 2D, since the sample is wide enough and the ignition is uniform across the fuel width.

Four cases, at 1 atm. 35% O_2 in N_2, at forced flows of 2, 5, 10, and 20 cm/s are presented here. This flow range captures flame spread from strictly upstream spread at 2 cm/s flow velocity (top row) to predominantly downstream spread by 20 cm/s (bottom row).

The surface color view images show that while the upstream leading edge of the flame is 2D, there are 3D soot structures occurring in the flame. At 2 and 5 cm/s the soot along the edges of the fuel is most pronounced, and at 10 cm/s the soot funnels downstream over the downstream fuel.

The IR images show the fairly 2D thermal structure of the pyrolyzing fuel surface upstream and downstream even up to 10 cm/s (blue to pink transition). At 2 and 5 cm/s the only downstream exothermic char oxidation is at the fuel edges. Exothermic char oxidation that is a clear indication of a separate downstream flame base is beginning to show at 10 cm/s with a slight saturated glow at the burnout, and by 20 cm/s a highly exothermic concurrent flame base is clearly present in the IR, showing two separated flames. The gas-phase sooting increases in intensity in the burnout gap as flow increases.

This work was jointly supported by a NASA/NEDO international agreement.

Keywords:
Microgravity; infrared; cellulose; flow

Reference

1. S. L. Olson, J. R. Lee, O. Fujita, M. Kikuchi, T. Kashiwagi, Quantitative infrared image analysis of simultaneous upstream and downstream microgravity flame spread over thermally-thin cellulose in low speed forced flow, 8th U.S. National Combustion Meeting, University of Utah, Salt Lake City, UT, May 19–22, 2013.

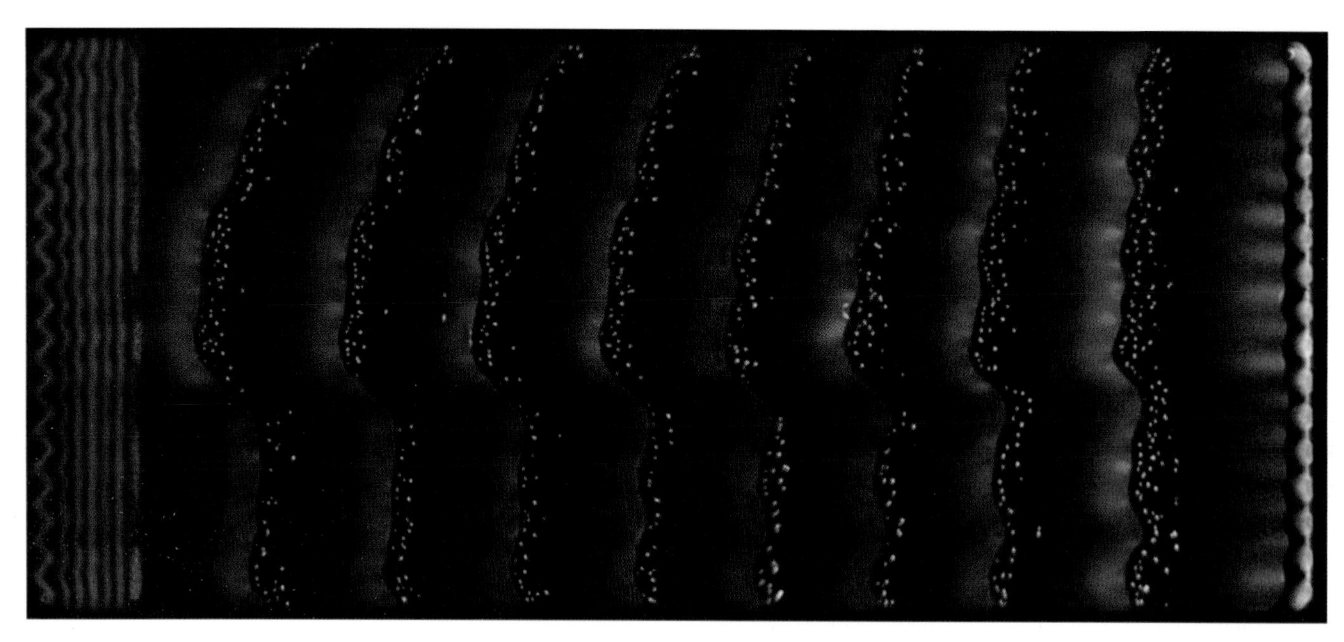

Figure 4.29

4.4.3 Large-Scale Flame Spread in Space

Paul Ferkul, Sandra Olson, David Urban, Gary Ruff, and John Easton
NASA

James T'ien and Ya-Ting Liao
Case Western Reserve University

Carlos Fernandez-Pello
University of California, Berkeley

Jose Torero
University of Maryland

Christian Eigenbrod
University of Bremen

Guillaume Legros
University Pierre et Marie Curie

N. Smirnov
Lomonosov Moscow State University

Osamu Fujita
Hokkaido University

Sebastien Rouvreau
Belisama R&D

Balazs Toth
ESA ESTEC

Grunde Jomaas
University of Edinburgh

NASA and an international team of investigators have worked to address open issues in spacecraft fire safety. NASA's Spacecraft Fire Safety Demonstration Project was developed with a goal to conduct a series of large-scale experiments in true confined spacecraft environments that represent practical spacecraft fires. One flight examined flame spread over a large thin sheet of flammable fuel (cotton/fiberglass 41 cm × 94 cm). This experiment was performed on an unmanned ISS resupply spacecraft after they had delivered their cargo and had undocked, but prior to the reentry. The opposed-flow experiment (left) was performed first, followed by the concurrent-flow experiment (right), shown in the above montage. The flow was 25 cm/s air flow from right to left. The image height is 41 cm, the full width of the sample.

The opposed-flow flame was ignited at the downstream end, and the flame was allowed to propagate upstream (left image sequence, images 10 s apart). The flame quickly reached a steady size and spread rate. The flame was very flat and uniform, and mostly blue. After a prescribed time, the flow was turned off, and the flame extinguished almost immediately. As shown in the last image of the left sequence above, the flame front broke up into smaller flamelets before extinguishing completely. It is noteworthy that in Earth's gravity, this fuel is not flammable in the downward configuration (opposed-flow).

For the concurrent ignition (rightmost image sequence, 40 s apart), the flame propagated on both sides of the igniter wire, with a string of small bright-blue upstream flame pearls and longer dim-blue downstream flames. Once the small piece of upstream fuel burned out, the concurrent flame developed a blue base and a pattern of orange-yellow radiation coming from incandescent soot. After the flame base passed, bright orange exothermic surface smolder spots burning some of the leftover fuel were visible.

The flame base was reasonably flat at first but developed into a wavy shape as it spread. This shape slowly became more exaggerated as time progressed, as two broad portions of the flame that had moved ahead tended to move even further ahead.

The flame speed and size for both tests were measured from the video. The flame and pyrolysis position traces show that the pyrolysis propagation rate was steady for most of the tests. For the concurrent-flow flame, after a transient that drives an overshoot in the pyrolysis length, that length reached a plateau value.

This experiment was funded by the Advanced Exploration Systems Division of the Human Exploration and Operations Directorate at NASA.

Keywords:
Spacecraft safety; flame spread; microgravity; ignition

Reference

1. G. Jomaas, J. L. Torero, C. Eigenbrod, J. Niehaus, S. L. Olson, P. V. Ferkul, et al., Fire safety in space – beyond flammability testing of small sample, *Acta Astronautica* 109 (2015), 208–216.

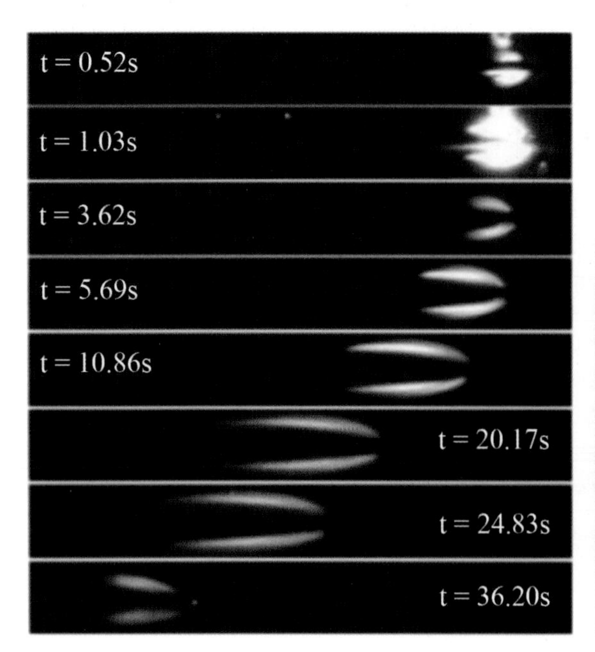

Figure 4.30

4.4.4 Concurrent Flame Growth, Spread, and Extinction over Composite Fabric Samples in Low-Speed Purely Forced Flow in Microgravity

Xiaoyang Zhao, Ya-Ting T. Liao, Michael C. Johnston, and James S. T'ien

Case Western Reserve University

Paul V. Ferkul and Sandra L. Olson

NASA Glenn

Flame spread over solids in concurrent-flow purely forced convection depends strongly on the magnitude of flow velocity. Theories predict that a limiting flame length will be reached and is a function of flow velocity.[1,2] Theory also predicts the flame will be quenched if the flow velocity is sufficiently low.[3] The figures describe an experiment (BASS) carried out on the International Space Station (ISS) within the Microgravity Science Glovebox. The fuel sample, consisting of a composite fabric (75% cotton, 25% fiberglass), is mounted on a sample holder. The exposed sample dimension is 10 cm × 2.2 cm. The flow velocities range from 1 to 55 cm/s. The side-view video images in the left figure show the sequence of flame growth, steady spread, and decay in air at a flow of 10 cm/s (blowing from right to left). After ignition, the flame length gradually increased, but at approximately 20 s, a limiting length was reached and stayed constant until 28 s when the flame tip reached the end of the sample. The spread rate during this 'quasi-steady' period was 2.75 mm/s. The right figure shows a quenching sequence at 18.7% oxygen (balanced with nitrogen), which occurred when the flow was reduced. These top view images of the flame were taken by a still camera. After ignition at 10 cm/s, the flow was turned down gradually until 2.2 cm/s was reached. This image shows that both the flame length and width became smaller until quenching was reached. The quenching velocity is a function of ambient oxygen percentage, among other parameters.

This work was supported by NASA.

Keywords:

Concurrent flame spread; forced flow; microgravity; limiting length; quenching; extinction

References

1. X. Zhao, Y.-T. T. Liao, M. C. Johnston, J. S. T'ien, P. V. Ferkul, S. L. Olson, Concurrent flame growth, spread and quenching over composite fabric samples in low speed purely forced flow in microgravity, *Proceedings of the Combustion Institute* 36 (2017), 2971–2978.
2. P. V. Ferkul, J. S. T'ien, A model of low-speed concurrent flow flame spread over a thin fuel, *Combustion Science and Technology* 99 (1994), 345–370.
3. Y.-T. Tseng, and J. S. T'ien, Limiting length, steady spread and nongrowing flames in concurrent flow over solids, *Journal of Heat Transfer* 132 (2010), 091201.

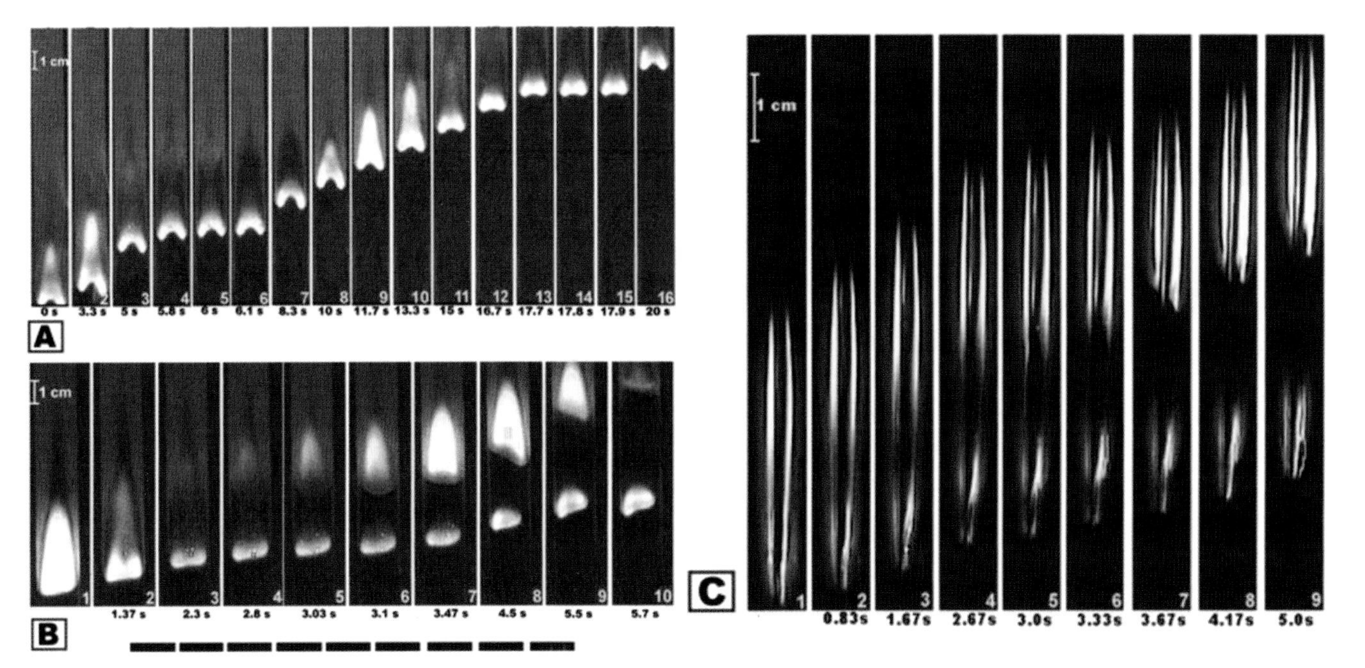

Figure 4.31

4.4.5 Cyclic Flame Growth Phenomena in Nomex III Fabric

Julie Kleinhenz
NASA Glenn

James Tien
Case Western Reserve University

Upward flame spread over samples of Nomex III fabric in enriched-oxygen, reduced-pressure environments resulted in a cyclic flame growth phenomena.[1] This is shown in the image sequences above. These images are digitally enhanced with increased contrast and brightness levels but no alteration to hue. (A) shows front (top) view images of the fuel surface. The flame lengthens, splits into two flames (image 4 and 5), and the secondary flame blows off immediately after break-off. This sequence then repeats. The conditions for this test were 29% oxygen, 6.25 psia, 2 cm width. (B) shows front-view images of a case where the secondary flame survives for several seconds. The conditions for this test were 30% oxygen, 6 psia, 2 cm wide. This was the second break-off of the test. Note that the pyrolysis front is visible in (A) and (B). (C) is a sequence of side (edge) view images showing a break-off similar to that in (B). The conditions for this test were 29% oxygen, 8 psia, 2 cm width. Based on TGA data and the measured surface temperatures during the test, this cyclic flame spread is believed to be the result of two-stage pyrolysis from the Nomex sample.

This work was funded by NASA.

Keywords:
Nomex flammability; upward flame spread limit; periodic upward flame spread; two; stage pyrolysis; flame splitting

Reference

1 J. Kleinhenz, J. S. T'ien, Combustion of Nomex® III fabric in potential space habitat atmospheres: cyclic flame spread phenomena, *Combustion Science and Technology* 179 (2007), 2153–2169.

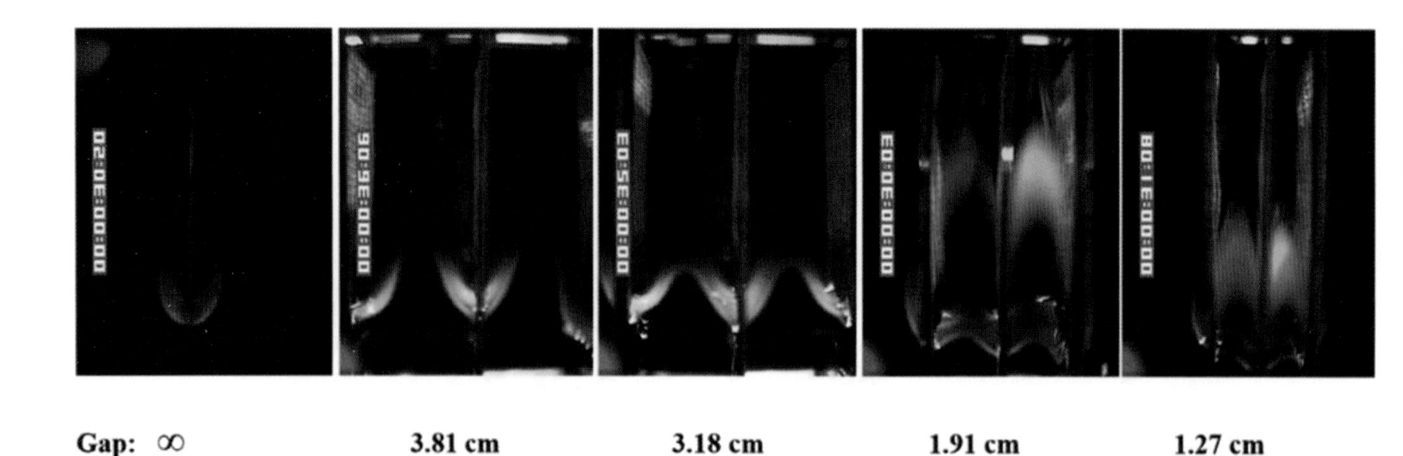

Figure 4.32

4.4.6 Concurrent Flame Spread over Three Parallel Thin Fuel Sheets in Microgravity

Sandra L. Olson

NASA Glenn

Concurrent flame spread experiments were performed in NASA Glenn Research Center's 5.18 s Zero Gravity Research Facility using the Microgravity Wind Tunnel Apparatus. The test conditions were selected to ensure radiation was important to the flame spread rate, so air at 5 cm/s concurrent flow was selected at 1 atm. total pressure.

Three 5 cm × 10 cm sheets of cheesecloth (half thickness area density 0.001 g/cm^2) were mounted on parallel metal frames. Three sheets were selected instead of only two so that the center sample would have a centerline plane of symmetry. The gap between the sheets was varied experimentally from 1.27 cm to 3.81 cm, and compared to a single sheet (infinite spacing). Three parallel igniter wires were used to ignite the three samples simultaneously after the experiment entered zero gravity. A green LED shines down from above the samples. Flow is up in each image.

The single sheet flame (∞ gap) under these conditions was all blue, with tips that are parallel to the fuel sample. At 3.81 cm spacing, the flame tips begin to interact, and the tips fan out away from the fuel surface. The flame is brighter, with a slight inner soot layer, which indicates that the flame temperature is higher than the single flame case. At 3.18 cm, the tips are connected but the flame bases are still well defined with a deep cleft between them. At 1.91 cm, the flames are connected through much of the body of the flame with a shallow cleft that is skewed toward the outer sheets. At 1.27 cm, the center flame base is lagging significantly behind the outer flames, and the cleft is shifted to the surface of the outer sheets. At these two narrowest gap spacings, a large vapor cloud of unburned fuel appears (illuminated by the green LED), indicating that the mixing of fuel and oxygen is inhibited for those conditions.

At low gap spacings, despite the enhanced radiant heat transfer with the higher radiation view factors, the heat release rate from the flame is significantly reduced due to limited available mixing of the fuel vapor and inflowing oxidizer in the narrow channels. At large gap spacings, the radiation exchange augments the flame's heat flux to the fuel surface, enhancing the burning. The optimum gap spacing appears to be close to 3.2 cm where the flame spread rate is increased to 152% of the single sheet value. These results are in qualitative agreement with a published numerical model.

This work was supported by NASA.

Keywords:
Microgravity; parallel sheets; gap spacing; flame spread; radiative exchange

References

1. S. L. Olson, Radiative exchange during concurrent flame spread in microgravity, Presented at the 2014 Spring Technical Meeting of the Central States Section of the Combustion Institute, Tulsa, OK, March 16–18, 2014.
2. H.-Y. Shih, Flame spread and interactions in an array of thin solids in low-speed concurrent flows, *Comb. Theory Modeling* 13, 3(2009), 443–459.

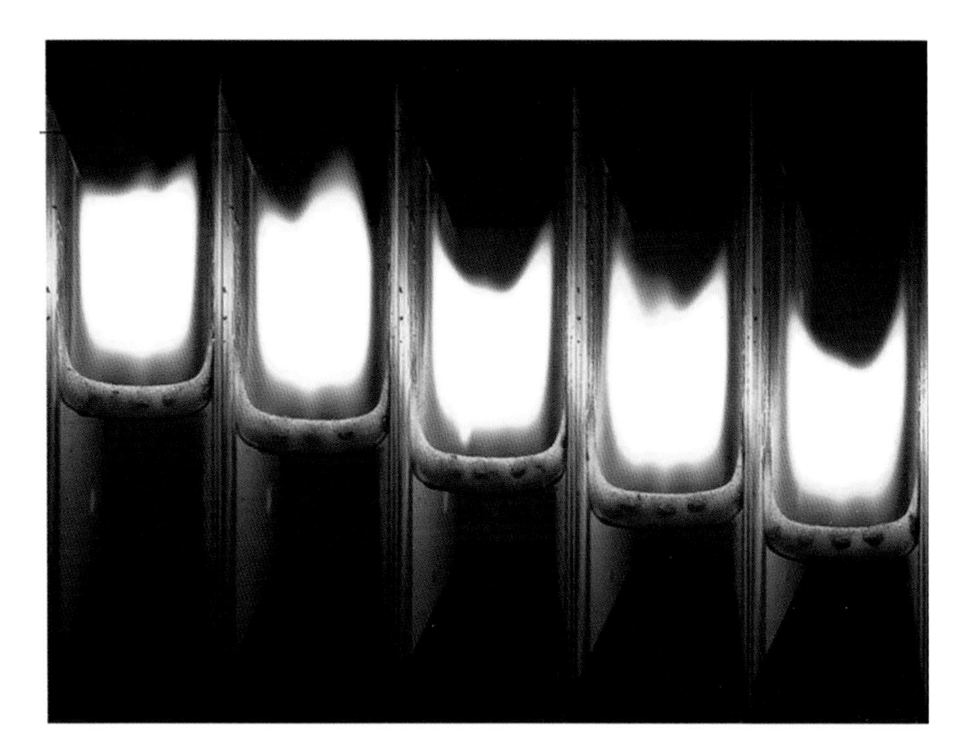

Figure 4.33

4.4.7 Diffusion Flame in Enclosed Geometry

Luca Carmignani and Subrata Bhattacharjee
San Diego State University

Keywords:
Flame spread; downward flame; solid fuel

A polymethyl-methacrylate (PMMA) sample with thickness of 3,175 μm and width of 30 mm is burning in ambient conditions between two ceramic plates. The enclosed geometry allows to approximate this diffusion flame as two-dimensional, with the sample width being the negligible coordinate. Moreover, enclosed flames generally show a laminar behavior also for fuel thicknesses that would generate turbulent flames, if the sample edges were free to burn.

The flame over the sample, as illustrated in the sequence of pictures, shows a steady-state behavior while spreading downward. The time step between each flame picture in the sequence is about 2 min, and the ignition transient is not considered. The burning area and flame length are constant throughout the experiment. Both characteristics depend on the sample thickness, and experiments show that the length of the melting front increases almost linearly with fuel thickness. The flame leading edge, characterized by the blue color, is spreading with a constant speed of 0.06 mm/s.

This work was supported by NASA.

References

1. S. Bhattacharjee, M. Bundy, C. Paolini, G. Patel, W. Tran, A novel apparatus for flame spread study, *Proceedings of the Combustion Institute* 34, 2 (2013), 2513–2521.
2. S. Bhattacharjee, M. Laue, L. Carmignani, P. V. Ferkul, S. L. Olson, Opposed-flow flame spread: a comparison of microgravity and normal gravity experiments to establish the thermal regime, *Fire Safety Journal* 79 (2016), 111–118.
3. S. Bhattacharjee, C. Paolini, W. Tran, J. R. Villaraza, S. Takahashi, Temperature and CO_2 fields of a downward spreading flame over thin cellulose: a comparison of experimental and computational results, *Proceedings of the Combustion Institute* 35, 3 (2015), 2665–2672.
4. S. Bhattacharjee, W. Tran, M. Laue, C. Paolini, Y. Nakamura, Experimental validation of a correlation capturing the boundary layer effect on spread rate in the kinetic regime of opposed-flow flame spread, *Proceedings of the Combustion Institute* 35, 3 (2015), 2631–2638.

Figure 4.34

4.4.8 Effect of a Varying Opposed-Flow Velocity in Microgravity

Luca Carmignani and Subrata Bhattacharjee
San Diego State University

Sandra L. Olson and Paul V. Ferkul
NASA Glenn

The Burning And Suppression of Solid – II (BASS-II) investigation considered different solid fuel geometries in presence of a concurrent or opposing flow field. Experiments were carried out inside the Microgravity Science Glovebox, on board the International Space Station.

Flames over thin solid fuels, exposed to an opposed flow, react quickly to velocity variations, therefore multiple velocities can be examined during the same experiment.

The sequence of flame pictures shows a diffusion flame spreading against a varying opposed-flow velocity, from 4 cm/s (first frame considered) to 0 cm/s (last frame), and at steps of 1 cm/s. The fuel is polymethyl-methacrylate (PMMA), with a thickness of 200 μm, burning with oxygen concentration of 20% and pressure of 1 atm.

From the sequence, it is clear that the yellow and orange parts of the flame tend to decrease with lower velocities. The flame is burning in the so-called radiative regime, where radiation is the most important heat transfer mechanism. This regime can only be achieved in microgravity, where buoyancy is absent and the oxidizer can reach the reaction zone mostly by diffusion. As a result, the contour of the flame is elliptical and characterized by a bright blue color typical of premixed flames, in contrast with the common tailored shape of flames in normal gravity.

This work was supported by NASA.

Keywords:
BASS–II; opposed-flow; flame spread

References

1. S. Bhattacharjee, A. Simsek, S. L. Olson, P. V. Ferkul, The critical flow velocity for radiative extinction in opposed-flow flame spread in a microgravity environment: a comparison of experimental, computational, and theoretical results, *Combustion and Flame* 163 (2016), 472–477.
2. S. Bhattacharjee, M. Laue, L. Carmignani, P. V. Ferkul, S. L. Olson, Opposed-flow flame spread: a comparison of microgravity and normal gravity experiments to establish the thermal regime, *Fire Safety Journal* 79 (2016), 111–118.
3. S. Bhattacharjee, A. Simsek, F. Miller, S. Olson, P. Ferkul, Radiative, thermal, and kinetic regimes of opposed-flow flame spread: A comparison between experiment and theory, *Proceedings of the Combustion Institute* 36 (2017), 2963–2969.

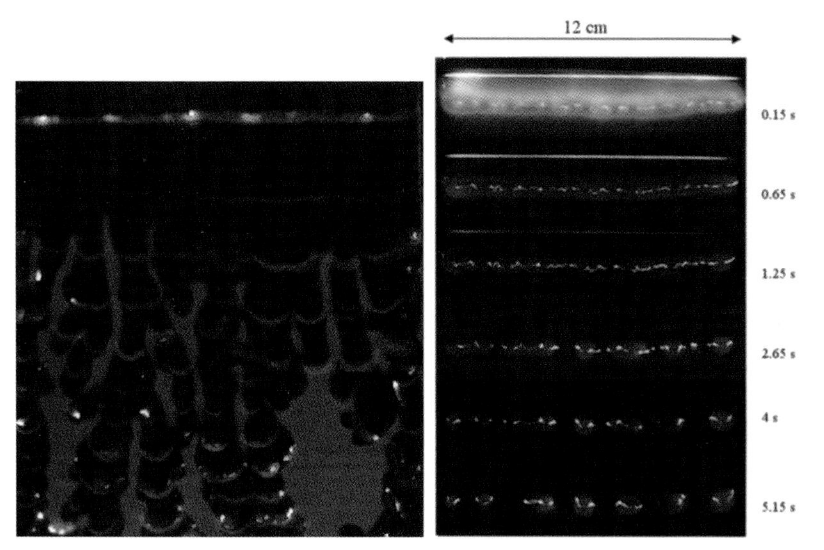

Figure 4.35

4.4.9 Fingering Flamelet Spread over Thin Cellulosic Fuels

Sandra L. Olson
NASA Glenn

Indrek S. Wichman
Michigan State University

Fletcher J. Miller
San Diego State University

Fingering flamelets are a near-limit fire adaptation to adverse conditions of limited oxidizer flow or diffusion. These adverse conditions occur in spacecraft or on earth in vertically confined spaces where there is no room for the hot gases to rise and entrain fresh air. In these near-limit conditions, a flame front propagating across a solid fuel breaks apart into separate flamelets.[1]

On the left image montage,[2] a 17.5 cm-wide Whatman 44 ashless filter paper is mounted in a narrow channel apparatus, spaced 7 mm from the top window and 5 mm from the lower copper plate. The initial forced airflow rate was 10 cm/s for ignition (first flame), then the flow was decelerated at 1.05 cm/s^2 to a flow velocity of 5 cm/s. The flame becomes corrugated (the second dim blue flame), and the flow is further decelerated at 0.1 cm/s^2 to a final flow velocity of 3 cm/s. The third flame front from the top shows breakup into flamelets. Thereafter, the flamelet fingering spread process is a sequence of bifurcations, extinctions, and recombinations. The unburned fuel left behind is illuminated by a red LED.

On the right sequence,[3] the same fuel is tested in the Zero Gravity Research Facility 5.18 s drop tower. The sequence of images shows the flame transition from an initial normal gravity uniform flame to a corrugated flame (1.25 s) to seven flamelets evenly spaced along the front that survived to the end of the drop test. The Whatman 44 filter paper sample is 12 cm wide. Time from the start of the drop is indicated on the right of the image. Test conditions were an air velocity of 8 cm/s with a copper substrate mounted 2.5 mm behind the sample.

Flamelets form a Turing-type reaction–diffusion fingering pattern as they spread across the fuel. A the continuous logistic model[4] with a time lag can describe the flamelet population growth and fluctuation around a stable population's carrying capacity based on environmental limitations.

This work was supported by NASA.

Keywords:
Opposed flame spread; fingering; flamelets; narrow channel apparatus

References

1. I. S. Wichman, S. L. Olson, F. J. Miller, A. Hariharan, Fire in microgravity, *American Scientist*, 104 (2016), 44–51.
2. I. S. Wichman, S. L. Olson, F. J. Miller, S. A. Tanaya, Experimental evaluation of flame and flamelet spread over cellulosic materials using the narrow channel apparatus, *Fire & Materials* 37 (2013), 503–519.
3. S. L. Olson, F. J. Miller, S. Jahangirian, I. S. Wichman, Flame spread over thin fuels in actual and simulated microgravity conditions, *Combustion and Flame* 156 (2009), 1214–1226.
4. S. L. Olson, F. J. Miller, I. S. Wichman, Characterizing fingering flamelets using the logistic model, *Combustion Theory and Modelling* 10, 2 (2006), 323–347.

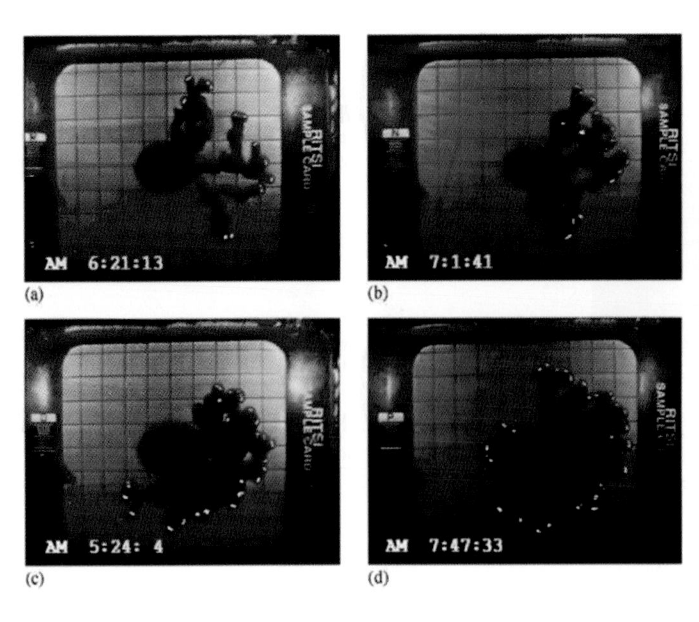

(e)

Figure 4.36

4.4.10 Microgravity Fingering Smolder Spread over a Thin Cellulosic Fuel

Sandra L. Olson
NASA Glenn

Howard Baum and Takashi Kashiwagi
NIST

Smoldering flame spread tests were conducted in the Middeck Glovebox Facility on the STS-75 USMP-3 space shuttle mission.[1] The Whatman 44 ashless filter paper was doped with smolder-promoting potassium acetate. A 1 cm grid was printed on each sample and a red LED illuminated the surface. Ignition occurred in the center of the sheet, and the smolder front propagated out from the center. A slow airflow was imposed from the right side of each image. The airflow velocities were: (a) 0.5 cm/s, (b) 2 cm/s, (c) 5 cm/s, and (d) 6.5 cm/s.

A complex finger-shaped char pattern was observed in microgravity. This is in contrast with a symmetric horizontal smolder spread in normal gravity with only buoyant airflow (~20 cm/s), shown in (e). For the microgravity fingering smolder spread, each "fingertip" had a glowing smolder front that propagated, frequently bifurcated, and occasionally extinguished. The smolder fingers preferentially propagated upstream (to the right) into fresh oxidizer. Downstream propagation was slower and less viable. At low imposed flows, onset of downstream smolder was delayed until completion of upstream smolder, due to the "oxygen shadow" cast by the upstream smolder fronts. The smolder fronts cumulatively follow population dynamics, and the population growth rate increases linearly with the forced flow velocity. The data indicate that a uniform smolder front in the upstream direction should occur when the forced flow velocity exceeds 9–10 cm/s.

Smolder front bifurcation occurred when the smolder front width exceeded approximately 0.3 cm. This is of the same order as the diffusive gas-phase Stokes length scale ($l_g = 2D/U_\infty$) and is significantly larger than the solid-phase length scale $l_s = \alpha_s / V_s$, where α_s is the solid thermal diffusivity and V_s is the velocity of the smolder spot. The bifurcation process occurs in approximately 1 s, regardless of imposed flow. This timescale is consistent with solid-phase timescales $t_s = l_s / V_s \approx 2/3$ s, because the primary heat release and heat losses are from the solid phase. In contrast, gas-phase timescales vary with flow $t_g = l_g/U_\infty \approx 0.8$ s to less than 0.1 s at the highest flows. This bifurcation-extinction process is consistent with the oxygen transport model.

This work was supported by NASA.

Keywords:
Smoldering; fingering

Reference

1. S. L. Olson, H. R. Baum, T. Kashiwagi, Finger-like smoldering over thin cellulosic sheets in microgravity, *Proceedings of the Combustion Institute* 27 (1998), 2525–2533.

5 Industrial Flames

Edited by Charles E. Baukal, Jr.

Introduction

Charles E. Baukal, Jr.

Industrial combustion processes are unique compared to the other chapter subjects in this book by virtue of their scale and diversity. For example, a large flame produced by a flare in a petrochemical plant could exceed 100 m in length, and individual burners in large power boilers could be over 100 MW firing capacity each. There are numerous types of combustors that have many names such as heaters, furnaces, ovens, kilns, dryers, boilers, thermal oxidizers, crackers, reformers, and smelters. The processes range in temperatures from as low as 300K for some food heating and drying processes to as high as over 2,000 K for some glass and metals production processes. Fuels can be solid (e.g., coal), liquid (e.g., oil), or gaseous (e.g., natural gas). Some processes are batch (e.g., rotary aluminum melting furnaces) or continuous (e.g., float glass furnaces). The combustors come in a wide range of shapes and sizes including both horizontal (e.g., cement kiln) and vertical cylinders (e.g., vertical cylindrical process heaters), approximately spheres (e.g., electrical arc furnaces with supplement burners), and rectangular boxes (e.g., cabin heaters used in the petrochemical industry). Some include heat recuperation that may be recuperative (e.g., air preheaters used on process heaters) or regenerative (e.g., air preheaters used in glass melting furnaces). Some processes use oxygen-enhanced combustion ranging from low-level enrichment (e.g., to enhance scrap aluminum melting) up to complete replacement of the combustion air with pure oxygen (e.g., supplemental burners used to enhance electric arc furnaces).

This chapter has been divided into six sections: metals industry, process heating, power generation, infrared heating and drying, flares, and oxygen-enhanced flames. Some of those sections are further broken down into subsections. The purpose of this chapter is not to attempt to be comprehensive, as that would be beyond the scope of the book, but rather to be representative of the many types of industrial combustion processes. This should give the interested reader a flavor for what combustion is used for in industry. As might be expected, heat transfer, thermal efficiency, pollution emissions, and reliability are important in most of these processes.

There are some good reference books on the subject of industrial combustion the interested reader could consult for further details on many of the processes pictured in this chapter. Griswold's (1946) book is very practically oriented and includes chapters on gas burners, oil burners, stokers and pulverized-coal burners, heat transfer, furnace refractories, tube heaters, process furnaces, and kilns [1]. Stambuleanu's (1976) book on industrial combustion has information on furnaces [2]. British Gas sponsored a very useful book on industrial combustion processes that specifically use natural gas as the fuel [3]. Trinks et al. (2004) have written what many consider to be the bible of industrial furnaces [4]. Deshmukh (2005) discusses a range of topics of interest in industrial combustion including fuels, burners, and refractories, where the emphasis is more on metal treatment [5]. Mullinger and Jenkins (2008) discuss a wide range of industrial furnaces and processes [6]. There is a series of books on industrial combustion that includes much information on a wide range of aspects including heat transfer [7], computational fluid dynamics [8], pollution [9], burners [10], testing [11], oxygen enhancement [12], and specific types of process in the petrochemical industry [13, 14].

References

1. J. Griswold, *Fuels, Combustion and Furnaces*, McGraw-Hill, New York, 1946.
2. A. Stambuleanu, *Flame Combustion Processes in Industry*, Abacus Press, Tunbridge Wells, UK, 1976.
3. J. R. Cornforth (ed.), *Combustion Engineering and Gas Utilisation*, 3rd ed., E&FN Spon, London, 1992.
4. W. Trinks, M. H. Mawhinney, R. A. Shannon, R. J. Reed, J. R. Garvey, *Industrial Furnaces*, 6th ed., John Wiley & Sons, New York, 2004.
5. Y. V. Deshmukh, *Industrial Heating*, CRC Press, Boca Raton, FL, 2005.
6. P. Mullinger, B. Jenkins. *Industrial and Process Furnaces*. Butterworth-Heinemann, Oxford, UK, 2008.
7. C. E. Baukal, *Heat Transfer in Industrial Combustion*, CRC Press, Boca Raton, FL, 2000.
8. C. E. Baukal, V. Y. Gershtein, X. M. Li (eds.), *Computational Fluid Dynamics in Industrial Combustion*, CRC Press, Boca Raton, FL, 2001.
9. C. E. Baukal, *Industrial Combustion Pollution and Control*, Marcel Dekker, New York, 2004.
10. C. E. Baukal (ed.), *Industrial Burners Handbook*, CRC Press, Boca Raton, FL, 2004.
11. C. E. Baukal (ed.), *Industrial Combustion Testing*, Taylor & Francis, Boca Raton, FL, 2011.
12. C. E. Baukal (ed.), *Oxygen-Enhanced Combustion*, Second Edition, CRC Press, Boca Raton, FL, 2013.
13. C. E. Baukal (ed.), *The John Zink Hamworthy Combustion Handbook*, 2nd ed., 3 Volumes, CRC Press, Boca Raton, FL, 2013.
14. S. Londerville and C. E. Baukal (eds.), *Coen Hamworthy Combustion Handbook*, CRC Press, Boca Raton, FL, 2013.

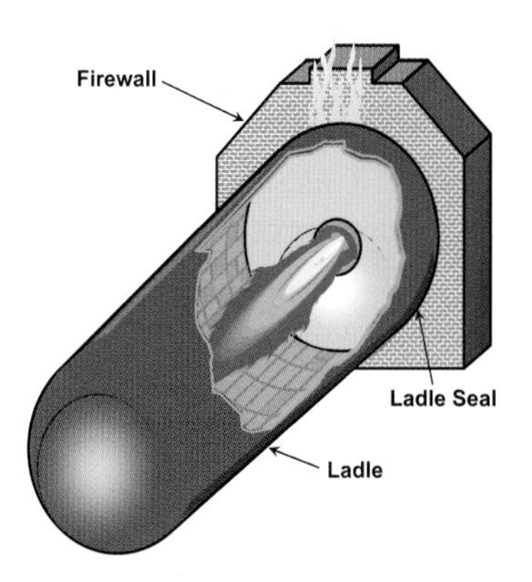

Figure 5.1 Drawing of a ladle used to transport molten metal which is being heated by a burner to keep the liquid metal from solidifying. Source: C. Baukal, "Introduction," in *Industrial Burners Handbook*, edited by C. E. Baukal, Jr., CRC Press, Boca Raton, FL, 2004

Figure 5.2 Actual ladle on a stand with a burner mounted on an endwall on the left with the burner firing from left to right. Source: C. E. Baukal, Jr., *Industrial Combustion Pollution and Control*, Marcel Dekker, New York, 2004

5.1 Metals Industry

5.1.1 Ladle Preheating

Nicholas Docquier, Michael Grant, and Kenneth Kaiser
Air Liquide

Charles E. Baukal, Jr.
John Zink Hamworthy Combustion

Figure 5.1 shows a schematic of a burner firing into a vessel called a *ladle* that is used to transport molten metal from one station to another in a metals production plant. Burners are used to keep the refractory lining of the ladle hot so the molten metal will not freeze and solidify inside the ladle and the refractory will not be thermally shocked. Figure 5.2 shows a burner mounted in an end wall firing into a ladle. Figure 5.3 shows a burner firing on an end wall of a ladle preheating station without a ladle in place so the flames can be seen. This burner uses both air and oxygen as the oxidizer. The amount of oxygen used depends on the size of the ladle, the temperature of the ladle, the cost of the oxygen and the fuel, and the turnaround time needed for the ladle. Figure 5.3 shows a 2 MW natural gas flame firing with (a) minimum oxygen participation, (b) medium oxygen participation, and (c) maximum oxygen participation.

Keywords:
Ladle preheating; oxygen-enhanced combustion; metals production

(a)

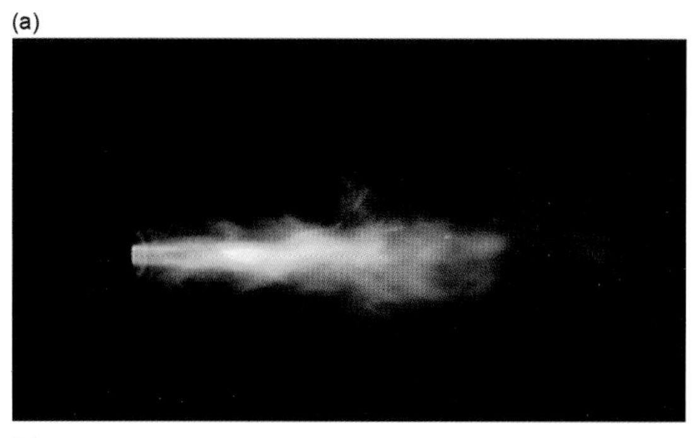

(b)

(c)

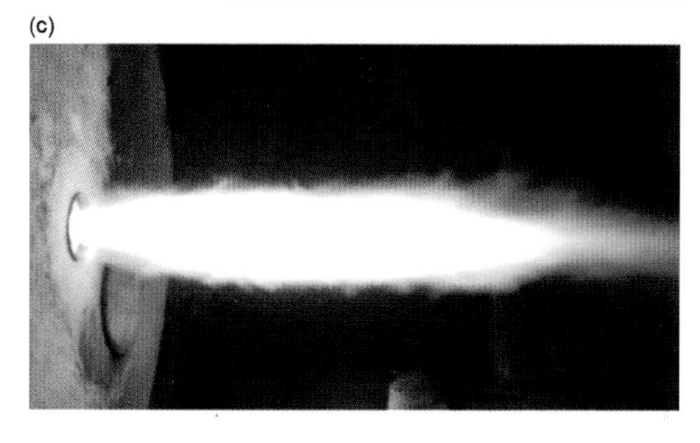

Figure 5.3 2 MW natural gas flame with (a) minimum oxygen enrichment of the combustion air, (b) medium oxygen enrichment, and (c) high-purity oxygen. Source: Nicolas Docquier, Michael Grant, and Kenneth Kaiser, "Ferrous Metals," Chapter 22 in *Oxygen-Enhanced Combustion*, edited by C. E. Baukal, Jr., CRC Press, Boca Raton, FL, 2013

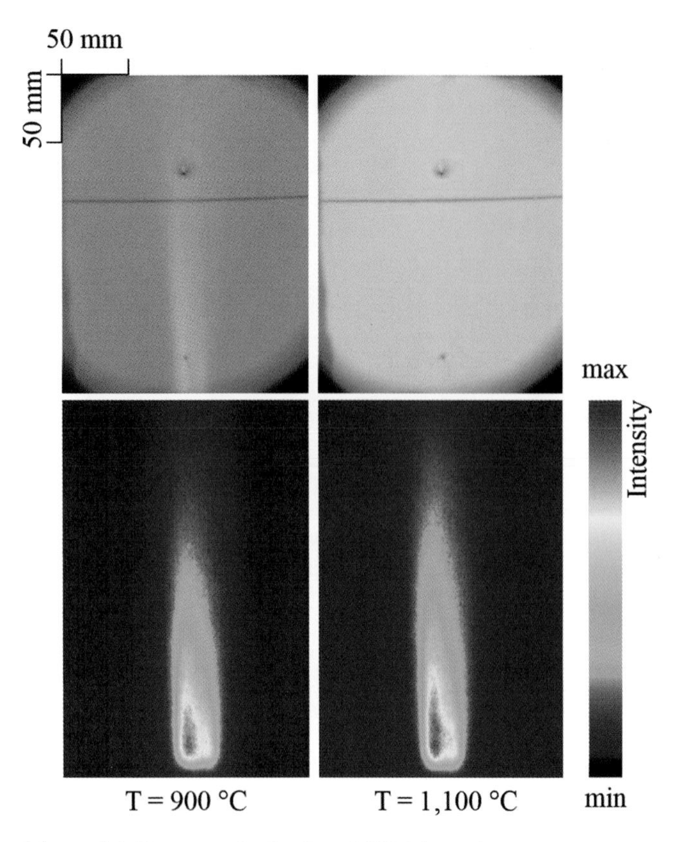

Figure 5.4 Photographs (top) and OH* intensity measurements for a recuperative burner used in metal reheating at furnace temperatures of 900°C and 1100°C. Source: C. Schwotzer, M. Schnitzler, H. Pfeifer, H. Ackermann, K. Lucka, "Experimental Investigation of a Concept for Scale Free Reheating of Semi-Finished Metal Products," Proceedings of the 7th European Combustion Meeting, 30 March - 2 April 2015, Budapest

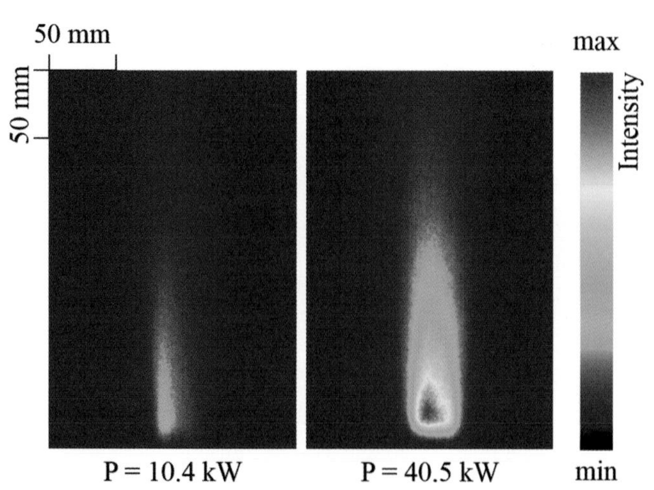

Figure 5.5 OH* intensity measurements for a recuperative burner used in metal reheating at 10.4 kW and 40.5 kW firing rates in a furnace at 900°C. Source: C. Schwotzer, M. Schnitzler, H. Pfeifer, H. Ackermann, K. Lucka, "Experimental Investigation of a Concept for Scale Free Reheating of Semi-Finished Metal Products," Proceedings of the 7th European Combustion Meeting, 30 March - 2 April 2015, Budapest

5.1.2 Experimental Investigation of a Concept of Low-Scale Reheating with Fuel-Rich Combustion

Christian Schwotzer and Herbert Pfeifer

RWTH Aachen University

Industrial furnaces for reheating semifinished metal products are often direct fired with natural gas and air. To ensure a complete combustion, the furnaces are fired fuel lean. Oxidation of the metals exposed to the furnace atmosphere causes significant material losses and additional work in further processing. A reheating concept, which reduces scale formation, was developed. It involves fuel-rich combustion, post-combustion of the unburned off-gas, and efficient preheating of the combustion air. Experiments show a significant reduction of scale formation for copper and copper-nickel alloys in the off-gas of a fuel-rich combustion with an air ratio of λ = 0.96 and a maximum temperature of 950°C, but also for steel, alloy 1.2367 in off-gas of a combustion with an air ratio of λ = 0.95 and a maximum temperature of 1,152°C.

As part of the project, the primary combustion is investigated with different measurement techniques. To characterize the combustion at different operating conditions, especially the dimension of the flame, a camera system with an intensified CCD-sensor and a band pass filter (peak transmission at 308 nm) for the detection of OH*-chemiluminescence is used and compared to photos of the flame with an exposure time of 1/60 s. A conventional recuperative burner with a maximum capacity of 40 kW is used, which is fired with natural gas and cold air. The combustion chamber has a height of 0.94 m and a diameter of 0.6 m.

Figure 5.4 shows a photo of the flame compared to the measured OH*-intensity at a temperature of 900°C and 1,100°C, a burner capacity of 30.5 kW, and an air ratio of λ = 1.15. The figure of the OH*-intensity is obtained by calculating the average of 300 exposures that were recorded in 30 seconds. Figure 5.5 shows the distribution of the OH*-radicals for an increasing burner capacity from 10.4 kW to 40.5 kW at an air ratio of λ = 1.15 and a furnace temperature of 900°C. Figure 5.6 shows the distribution of the OH*-chemiluminescence at different air ratios at a furnace temperature of 1,050°C.

The results show a stable combustion for a burner capacity of 50–100% with an air ratio of 1.15–0.7, respectively. The burner capacity of 50% was limited due to soot formation. For the fuel-rich combustion the measurements of the OH*-distribution show a decreasing length of the reaction zone with increasing air ratio, whereas the position of the maximum intensity remains constant.

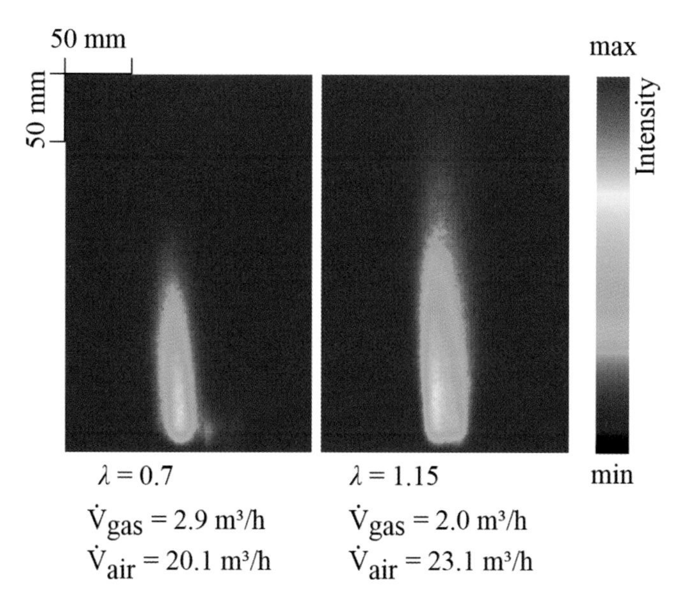

Figure 5.6 OH* measurements at different air/fuel ratios for a recuperative burner used in metal reheating in a furnace at 1050°C. Source: C. Schwotzer, M. Schnitzler, H. Pfeifer, H. Ackermann, K. Lucka, "Experimental Investigation of a Concept for Scale Free Reheating of Semi-Finished Metal Products," Proceedings of the 7th European Combustion Meeting, 30 March - 2 April 2015, Budapest

Acknowledgments

The project of the Research Association for the Mechanical Engineering Industry e. V. (FKM) proposed by the Research Association for Industrial Thermoprocessing Equipment e. V. (FOGI) was promoted via the German Federation of Industrial Research Associations e. V. (AiF, AiF-No. 17810 N) as part of the program for the promotion of Industrial Collective Research (IGF) by the Federal Ministry for Economic Affairs and Energy on the basis of a decision by the German Bundestag.

Keywords:

OH*-chemiluminescence; fuel-rich combustion; low-scale reheating

References

1. C. Schwotzer, T. Balkenhol, M. Schnitzler, H. Pfeifer, H. Ackermann, K. Lucka, Experimental investigation of a concept for scale free reheating of semi-finished metal products. Proceedings of the 10th European Conference on Industrial Furnaces and Boilers, Porto Gaia, Portugal, April 7–10, 2015.
2. C. Schwotzer, M. Schnitzler, H. Pfeifer, Low scale reheating with recuperative burners, *Gaswärme International* 65, 3 (2016), 67–72 (in German).
3. C. Schwotzer, M. Schnitzler, H. Pfeifer, H. Ackermann, D. Diarra, Low scale reheating of semi-finished metal products in furnaces with a central recuperator, *Heat Processing* 14, 3 (2016), 83–89.

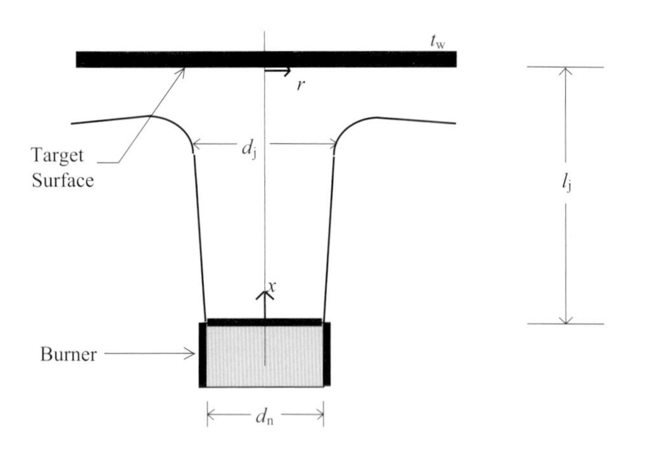

Figure 5.7 2D schematic of a flame jet impinging normally onto a plane surface. Source: C.E. Baukal, "Flame Impingement Measurements," Chapter 9 in *Industrial Combustion Testing*, edited by C.E. Baukal, CRC Press, Boca Raton, FL, 2011

Figure 5.8 Outlet of a Nordsea burner designed to fire high-purity oxygen with natural gas for use in metal heating. Source: C.E. Baukal, "Flame Impingement Measurements," Chapter 9 in *Industrial Combustion Testing*, edited by C.E. Baukal, CRC Press, Boca Raton, FL, 2011

(a)

(b)

Figure 5.9 Burner in Figure 5.8 firing natural gas with (a) air and (b) high-purity oxygen. Source: C.E. Baukal, "Flame Impingement Measurements," Chapter 9 in *Industrial Combustion Testing*, edited by C.E. Baukal, CRC Press, Boca Raton, FL, 2011

5.1.3 Flame Impingement

Charles E. Baukal, Jr.
John Zink Hamworthy Combustion

Direct flame impingement is used in the metals industry for reheating metals before they are shaped and formed. An example application is reheating the edge of a continuous metal sheet because the edges cool off more quickly than the center. The more nonuniform the temperature of the sheet, the more difficult it is to handle and process the metal, which often means either lower quality or more waste if the edges are cut off. Figure 5.7 shows a schematic of a flame impinging normal to a plane surface. A series of experiments were conducted using a commercial burner shown in Figure 5.8 made by Nordsea designed to using natural gas and high-purity oxygen. Figure 5.9 shows the effects of the oxidant composition with 5 kW natural gas flames using the following oxidants: (a) air and (b) high-purity oxygen. Figure 5.10 shows the effect of firing rate with oxygen/natural gas flames at firing rates of (a) 15 kW and (b) 35 kW. Figure 5.11 shows the effect of the distance between the burner and the target where (a) shows a close spacing and (b) shows a longer spacing.

This research was partially funded by the Gas Research Institute (Chicago, IL) for the project entitled "Development of Rapid-Heating Furnaces for Metal Reheating."

(a)

(b)

Figure 5.10 Burner in Figure 5.8 firing natural gas with high-purity oxygen at firing rates of (a) 15 kW and (b) 35 kW. Source: C.E. Baukal, "Flame Impingement Measurements," Chapter 9 in *Industrial Combustion Testing*, edited by C.E. Baukal, CRC Press, Boca Raton, FL, 2011

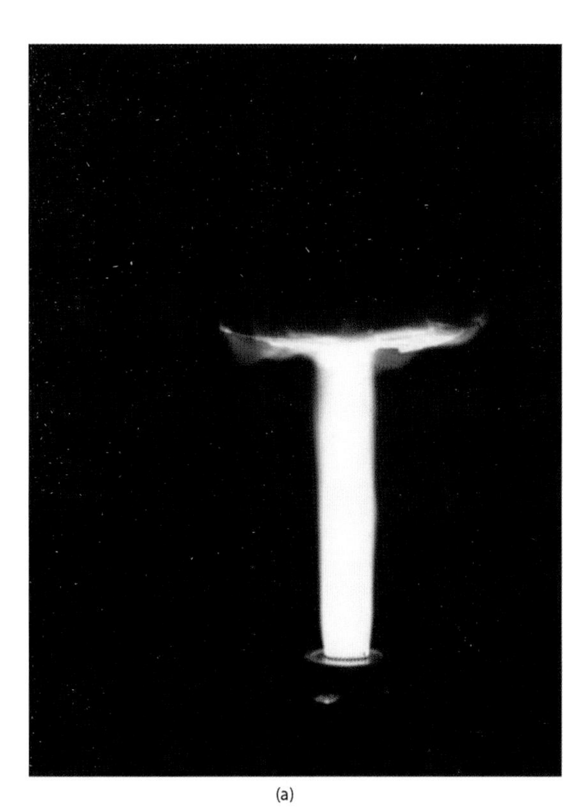

(a)

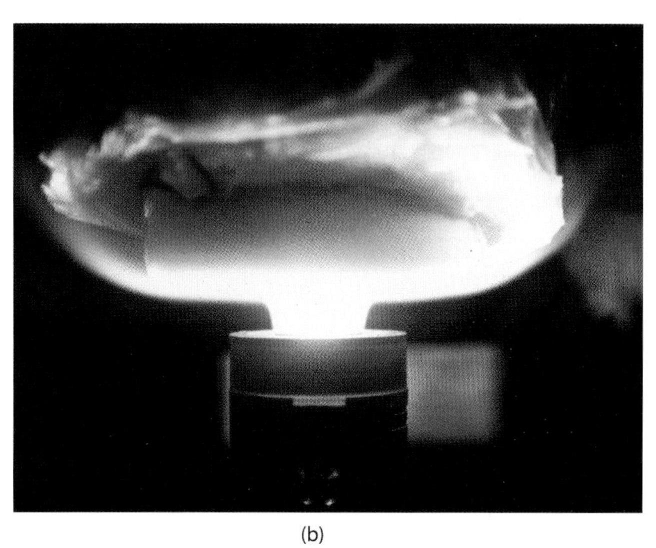

(b)

Figure 5.11 Burner in Figure 5.8 firing natural gas with high-purity oxygen with (a) long and (b) short spacing between the burner and the plane target surface. Source: C.E. Baukal, "Flame Impingement Measurements," Chapter 9 in *Industrial Combustion Testing*, edited by C.E. Baukal, CRC Press, Boca Raton, FL, 2011

Keywords:

Flame impingement; oxygen-enhanced combustion

References

1. C. E. Baukal, B. Gebhart, Heat transfer from oxygen-enhanced/natural gas flames impinging normal to a plane surface, *Experimental Thermal & Fluid Science* 16, 3 (1998), 247–259.
2. C. E. Baukal, *Heat Transfer from Flame Impingement Normal to a Plane Surface*, VDM Verlag, Saarbrücken, Germany, 2009.

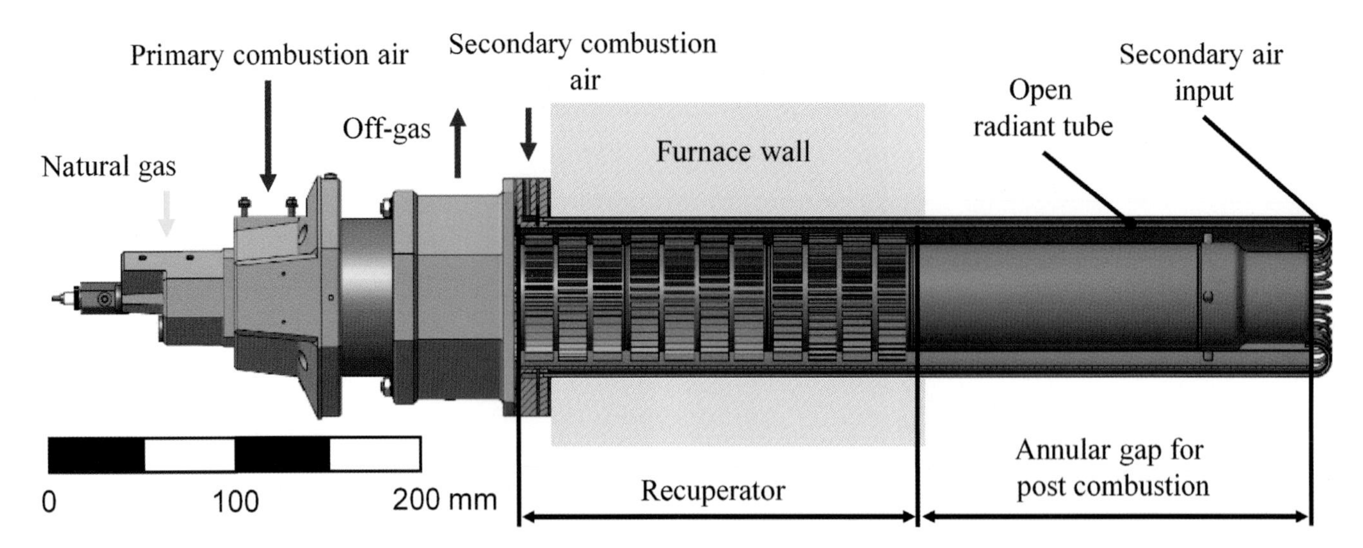

Figure 5.12 Drawing of a concept burner installed in a heat treating furnace. Source: N. Schmitz, C. Schwotzer, H. Pfeifer, J. Schneider, E. Cresci, J. G. Wünning, "Development of an Energy-Efficient Burner for Heat Treatment Furnaces with a Reducing Gas Atmosphere," *J. Heat Treatm. Mat.* 72 (2017), No. 2, pp. 73–80

5.1.4 Development of an Energy-Efficient Burner for Heat Treatment Furnaces with a Reducing Gas Atmosphere

Nico Schmitz, Christian Schwotzer, and Herbert Pfeifer
RWTH Aachen University

Joachim G. Wünning
WS Wärmeprozesstechnik GmbH

A main reason for metal loss of semifinished metal products during heating in reheating and heat treatment furnaces is scale formation. In this project, a burner was developed that produces a low oxidizing (reducing) atmosphere in the furnace. A recuperative burner generates a reducing furnace atmosphere due to fuel-rich combustion of natural gas and air. Complete combustion of the furnace atmosphere is ensured by injection of additional air and takes places in an open radiant tube resulting in high process energy efficiency.

Experimental and numerical methods were used to quantify operating conditions for the prototype burner design. The experimental setup consisted of a concept-burner (see Figure 5.12), which is installed in a combustion chamber. In the experiments, off-gas composition and temperature in the annular gap and in the off-gas after the recuperator were measured to determine the maximum temperature in the annular gap, length of the post-combustion zone, off-gas emissions, and energy efficiency of the burner.

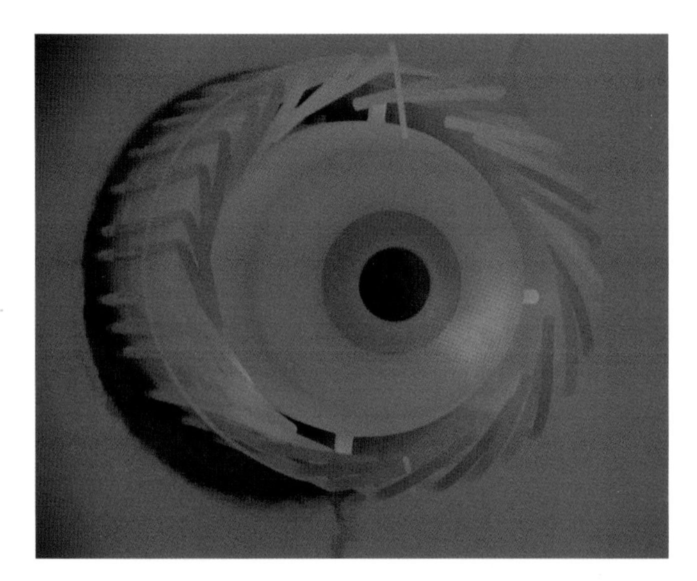

Figure 5.13 Burner shown in Figure 5.12 operating in flameless mode. Source: N. Schmitz, C. Schwotzer, H. Pfeifer, J. Schneider, E. Cresci, J. G. Wünning, "Development of an Energy-Efficient Burner for Heat Treatment Furnaces with a Reducing Gas Atmosphere," *J. Heat Treatm. Mat.* 72 (2017), No. 2, pp. 73–80

The burner operates in a flame and a flameless combustion mode. Figure 5.13 shows the burner in the flameless combustion mode, which is fired with natural gas and

Figure 5.14 Burner shown in Figure 5.12 operating in the visible flame mode with secondary air injection. Source: N. Schmitz, C. Schwotzer, H. Pfeifer, J. Schneider, E. Cresci, J. G. Wünning, "Development of an Energy-Efficient Burner for Heat Treatment Furnaces with a Reducing Gas Atmosphere," *J. Heat Treatm. Mat.* 72 (2017), No. 2, pp. 73–80

Figure 5.15 Burner shown in Figure 5.14 operating in the flameless mode. Source: N. Schmitz, C. Schwotzer, H. Pfeifer, J. Schneider, E. Cresci, J. G. Wünning, "Development of an Energy-Efficient Burner for Heat Treatment Furnaces with a Reducing Gas Atmosphere," *J. Heat Treatm. Mat.* 72 (2017), No. 2, pp. 73–80

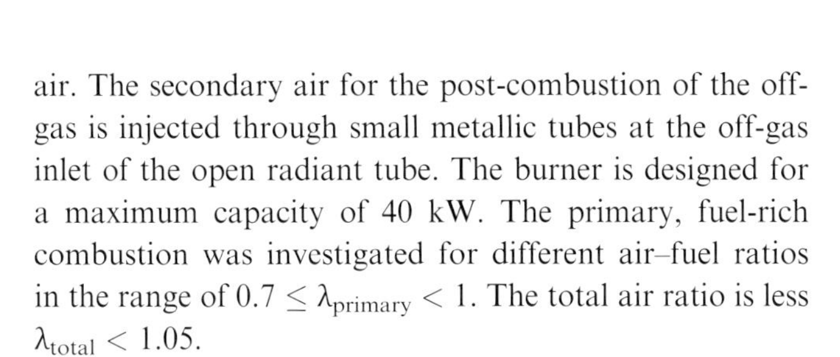

air. The secondary air for the post-combustion of the off-gas is injected through small metallic tubes at the off-gas inlet of the open radiant tube. The burner is designed for a maximum capacity of 40 kW. The primary, fuel-rich combustion was investigated for different air–fuel ratios in the range of $0.7 \leq \lambda_{primary} < 1$. The total air ratio is less $\lambda_{total} < 1.05$.

Figure 5.14 shows the burner firing in the flame mode. Secondary air is injected inside the annular gap. Therefore, there is no need for metallic tubes outside of the open radiant tube. The nominal burner capacity was 80 kW. The burner operated with a primary air ratio of $\lambda_{primary} = 0.75$ and a total air ratio of $\lambda_{total} \approx 1.15$, depending on the fuel and oxidizer. Due to the combination of fuel-rich combustion and post-combustion, the total NO_X emissions were low. For further reduction and temperature homogenization, the burner can operate in the flameless combustion mode (see Figure 5.15).

This project is supported in the Central Innovation Programme for SMEs (ZIM) by the Federal Ministry for Economic Affairs and Energy on the basis of a decision by the German Bundestag.

Keywords:

Recuperative burner; direct-fired furnace; fuel-rich combustion; post-combustion

References

1. N. Schmitz, J. Schneider, E. Cresci, C. Schwotzer, H. Pfeifer, J. G. Wünning, Development of an energy-efficient burner for heat treatment furnaces with a reducing gas atmosphere, 6th International Conference on Hot Sheet Metal Forming of High-Performance Steel, CHS^2–2017, Atlanta, GA, June 4–7, 2017.
2. N. Schmitz, C. Schwotzer, H. Pfeifer, J. Schneider, E. Cresci, J. G. Wünning, Numerical investigation on post-combustion in a burner for heat treatment furnaces with a reducing gas atmosphere. Proceedings of the 11th European Conference on Industrial Furnaces and Boilers, Albufeira, Portugal, April 18–21, 2017.

Figure 5.16 Radiant tube burner used in metal heating firing at 7.5 kW. Source: RWTH Aachen University

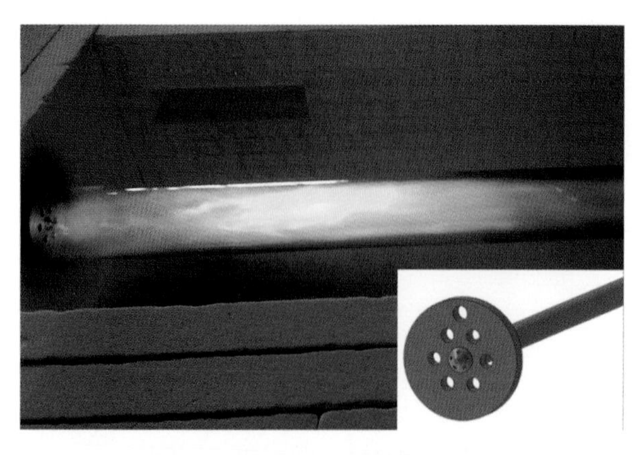

Figure 5.17 Modified radiant tube burner firing at 10 kW. Source: RWTH Aachen University

5.1.5 Experimental Investigation of the Combustion in Small Radiant Tubes for Multi-chamber Furnaces

Christian Schwotzer, and Herbert Pfeifer
RWTH Aachen University

The aim of this research-and-development project was an increase in efficiency and profitability of multi-chamber furnaces for press hardening. Therefore, an innovative furnace concept was developed, which is heated with small gas-fired radiant tubes. The concept consists of a modified multi-chamber furnace, which increases flexibility and productivity and reduces resource and energy consumption. Due to a compact design of furnace and heating components, less floor space is needed. Key components are small radiant tubes, which are fired with natural gas and air. These radiant tubes are used instead of conventional electric heating units.

Experimental investigations were done in a test combustion chamber with small radiant tubes made of glass. The internal dimensions of the combustion chamber were 480 mm × 300 mm × 1,630 mm (width × height × depth), with an insulation of 200 mm. The radiant tubes had a maximum diameter of 80 mm and a length of approximately 2,200 mm. The radiant tubes were heated with natural gas and cold air, which were fed into the radiant tube through a test burner. Different test burners were investigated. The focus of the investigation was on the stability of the combustion, the off-gas emissions, and the temperature profile at the surface of the radiant tube. Therefore, off-analysis and thermocouples are used.

Figure 5.16 shows a photo of the combustion in the radiant tube with a burner capacity of 7.5 kW and an air ratio of $\lambda = 1.05$. In further investigations, the burner capacity was increased up to 15 kW. The air ratio was < 1.09. Due to a gradient in temperature over the length of the radiant tube, the burner configuration was modified with numerical methods.

Figure 5.18 2 burners shown in Figure 5.17 firing in opposite directions inside a combustion chamber. Source: RWTH Aachen University

Figure 5.17 shows a photo of the combustion in the radiant tube with a modified test burner, based on numerical simulations, with a burner capacity of 10 kW and an air ratio of 1.08. In further investigations, the burner capacity was increased up to 15 kW. The air ratio was < 1.08. For more reduction of the CO emissions, the air ratio was increased to 1.25. The results showed changing combustion characteristics, but the temperature gradient was still in the same range.

To improve temperature distribution in the furnace, two radiant tubes were installed in the combustion chamber, with burners firing in opposite directions (Figure 5.18). The burners in this case were fired with a capacity of 15 kW each and an air ratio of 1.09. In -this test configuration, the temperature gradient was reduced from approximately 70 K to approximately 30 K.

This project was supported in the Central Innovation Programme for SMEs (ZIM) by the Federal Ministry for Economic Affairs and Energy on the basis of a decision by the German Bundestag.

Keywords:
Multi-chamber furnace; small radiant tubes; gas-fired burner

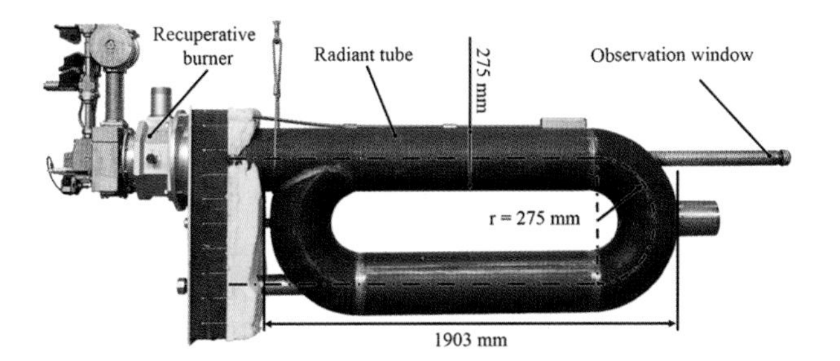

Figure 5.19 Experimental setup for a test of recuperative recirculating radiant tube burners used in metal heating applications. Source: N. Schmitz, C. Schwotzer, H. Pfeifer, "Measures to Improve Metallic Recirculating Radiant Tube Lifetime," *Gaswärme International*, 66 (2017), No. 4 (in German)

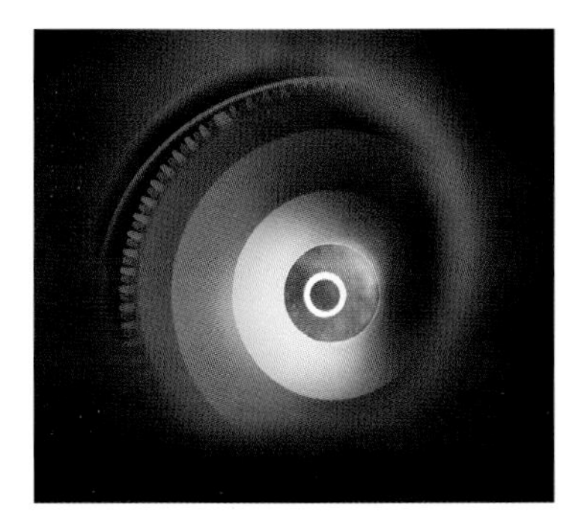

Figure 5.20 Outlet of a recuperative recirculating burner operating in the visible flame mode. Source: N. Schmitz, C. Schwotzer, H. Pfeifer, "Measures to Improve Metallic Recirculating Radiant Tube Lifetime," *Gaswärme International*, 66 (2017), No. 4 (in German)

5.1.6 Experimental and Numerical Investigation to Improve the Lifetime of Metallic Recirculating Radiant Tubes

Nico Schmitz, Christian Schwotzer, and Herbert Pfeifer

RWTH Aachen University

Joachim G. Wünning

WS Wärmeprozesstechnik GmbH

Radiant tubes operate at a high thermal and mechanical load. A premature failure of the tubes reduces the heat input to the furnace and therefore reduces productivity. The worst case would be the stop of the furnace. Besides choosing the appropriate material and a good dimensioning of the burner–radiant tube system, different influences lower the lifetime of radiant tubes.

In this project, some aspects of tube design were investigated using modern computational methods. These tools are very valuable to gain a better understanding of the complex mechanisms that can lead to premature tube failures. However, so far it is not possible to make exact predictions of tube life. There are still many phenomena, especially regarding high temperature creeping, which cannot be completely described. The other difficulty is that

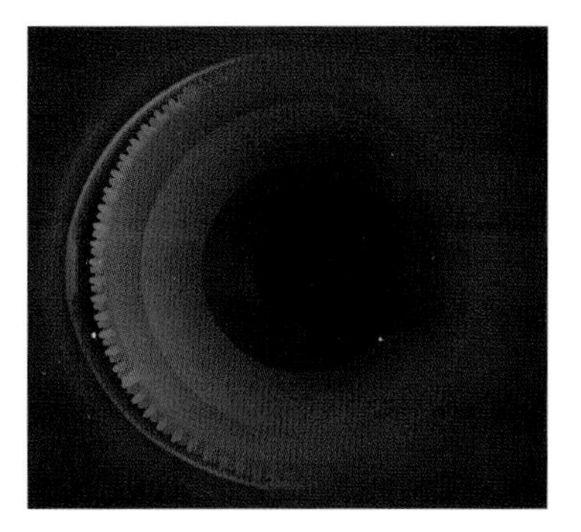

Figure 5.21 Outlet of a recuperative recirculating burner operating in the flameless mode. Source: N. Schmitz, C. Schwotzer, H. Pfeifer, "Measures to Improve Metallic Recirculating Radiant Tube Lifetime," *Gaswärme International*, 66 (2017), No. 4 (in German)

the boundary conditions of real operating data are largely unknown. However, despite its limitations, it can be very fruitful for suppliers and operators to cooperate with researchers to get a better understanding of their processes.

Figure 5.19 shows the experimental setup. For the experimental investigations and validation of the numerical models, a p-type test radiant tube was used, which was installed in an air-cooled combustion chamber. The radiant tube was fired with a recuperative burner fed with natural gas and air with a capacity of 120 kW and an air ratio of approx. $\lambda = 1.05$. The radiant tube was equipped with an observation window to characterize the combustion.

Figure 5.20 shows the outlet of the burner. The burner operates in the flame mode. There is a visible blue flame at the outlet of the combustion chamber of the burner. To minimize NO_X, emissions the burner can operate in a flameless combustion mode.

Figure 5.21 shows the outlet of the burner operating in the flameless combustion mode. In this case, there is no visible flame at the outlet of the burner; instead, the combustion reactions take place in a large reaction zone.

The project of the Research Association for the Mechanical Engineering Industry e. V. (FKM) proposed by the Research Association for Industrial Thermoprocessing Equipment e. V. (FOGI) was promoted via the German Federation of Industrial Research Associations e. V. (AiF, AiF-No. 17840 N) as part of the program for the promotion of Industrial Collective Research (IGF) by the Federal Ministry for Economic Affairs and Energy on the basis of a decision by the German Bundestag.

Keywords:
Radiant tube; flameless combustion; gas burner

Reference

1. N. Schmitz, M. Hellenkamp, H. Pfeifer, E. Cresci, J.G. Wünning, M. Schönfelder: Radiant tube life improvement for vertical galvanizing lines, 10th International Conference on Zinc and Zinc Alloy Coated Steel, AIST Galvatech, Toronto, Canada, May 31 – June 4, 2015.

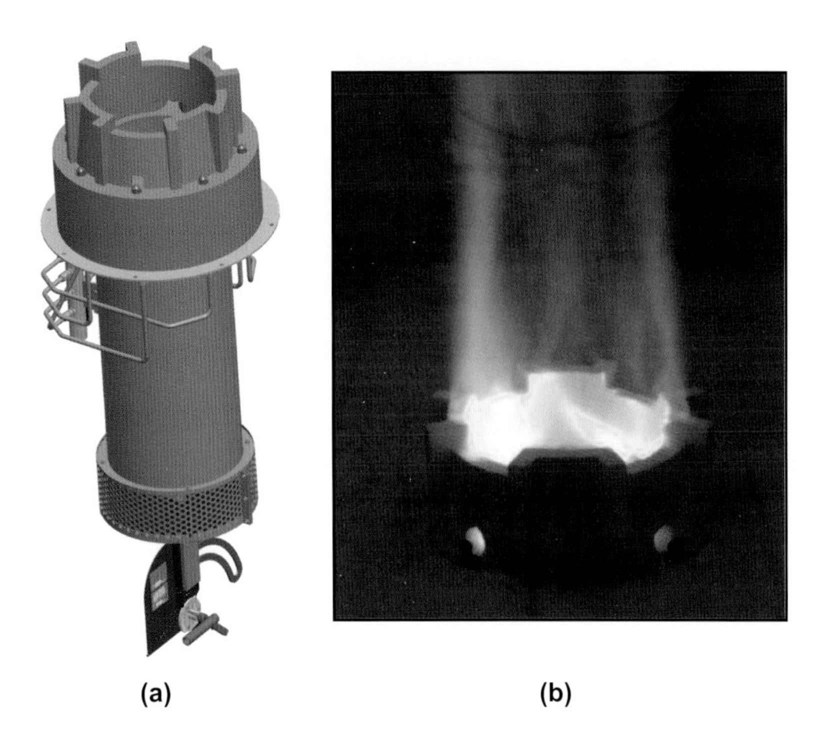

Figure 5.22 John Zink Hamworthy Combustion COOLstar® burner designed for ultra-low-NOx with a round flame shape. Source: John Zink Hamworthy Combustion (Tulsa, OK)

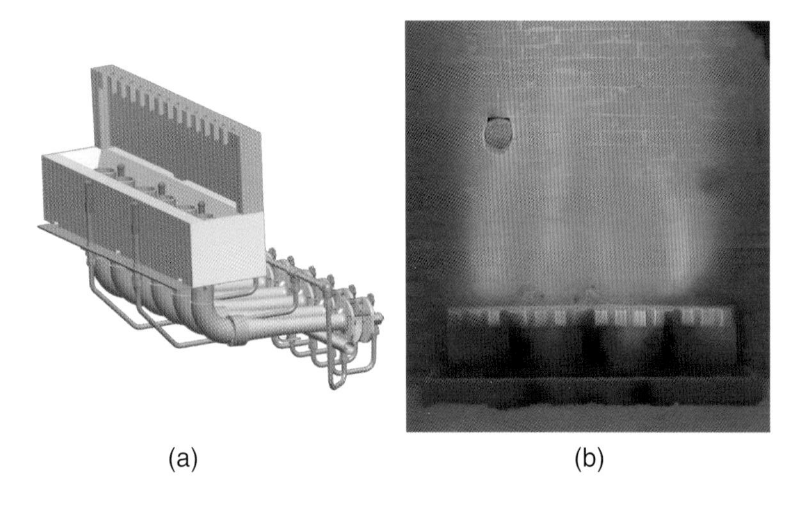

Figure 5.23 John Zink Hamworthy Combustion LPMF burner designed for ultra-low-NOx with a rectangular ("flat") flame shape. Source: C. E. Baukal, Jr. (ed.), *The John Zink Hamworthy Combustion Handbook*, Volume 3: Applications, CRC Press, Boca Raton, FL, 2013

5.2 Process Heating

5.2.1 Typical Burners

5.2.1.1 Typical Process Burner Flame Shapes

Eric Gebhard and Charles E. Baukal, Jr.
John Zink Hamworthy Combustion

In the process heating industry, by far the two most common flame shapes are round and rectangular. Figure 5.22a shows a drawing of a COOLstar® burner designed to produce a round flame shown in Figure 5.22b. Figure 5.23 shows a John Zink Hamworthy Combustion LPMF burner designed to produce a rectangular, sometimes referred to as a flat, shape flame as seen in Figure 5.23b.

Keywords:
Process burners; flame shape

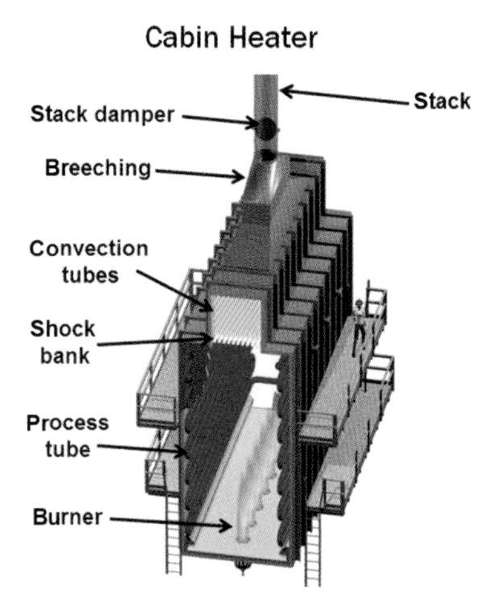

Figure 5.24a Cabin-style fired heater: (a) drawing. Source: John Zink Hamworthy Combustion (Tulsa, OK)

Figure 5.24b Cabin-style fired heater: (b) photograph. Source: C. E. Baukal, Jr. (ed.), *The John Zink Hamworthy Combustion Handbook*, Volume 3: Applications, CRC Press, Boca Raton, FL, 2013

Figure 5.25 Row of round burners firing in a cabin-style fired heater. Source: C. E. Baukal, Jr. (ed.), *The John Zink Hamworthy Combustion Handbook*, Volume 3: Applications, CRC Press, Boca Raton, FL, 2013.

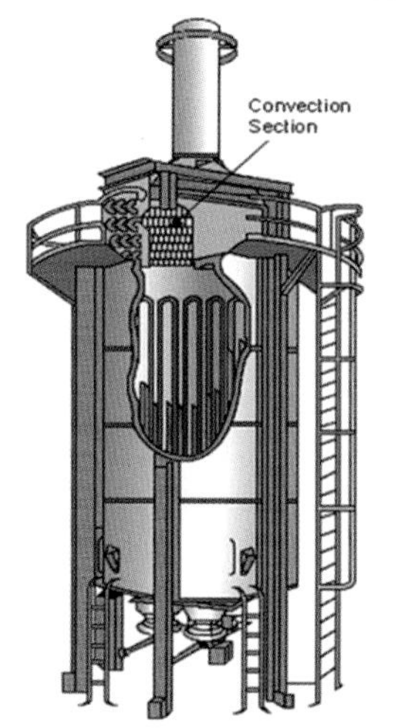

Figure 5.26 Vertical cylindrical fired heater. Source: John Zink Hamworthy Combustion (Tulsa, OK)

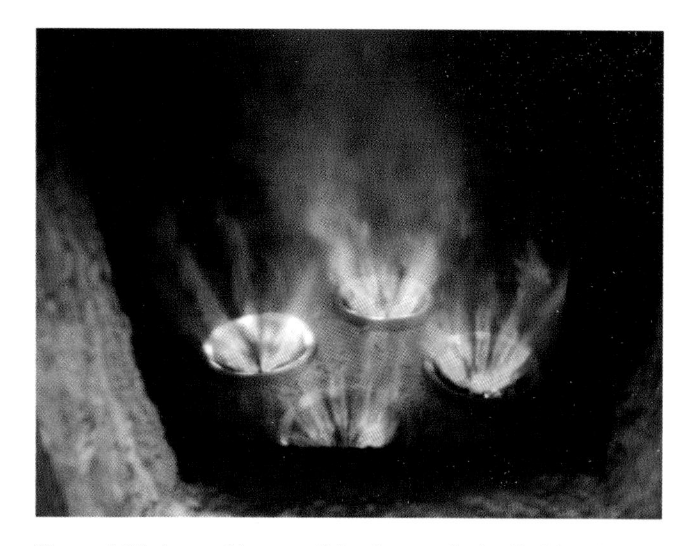

Figure 5.27 4 round burners firing in a vertical cylindrical heater. Source: John Zink Hamworthy Combustion (Tulsa, OK)

5.2.1.2 *Freestanding Burners*

Mike Pappe and Bill Johnson
John Zink Hamworthy Combustion

Freestanding burners are located away from the walls and generally fire upwardly. Two of the most common examples are shown here. Figure 5.24 shows a sketch and a photo of a cabin heater where burners are aligned in one or more straight rows. Figure 5.25 shows flames in a crude unit where the burners are firing on refinery fuel gas. Figure 5.26 shows a drawing and photo of a vertical cylindrical (VC) heater. Figure 5.27 shows flames from four burners in a circle firing refinery fuel gas in a VC heater.

Keywords:
Freestanding burners

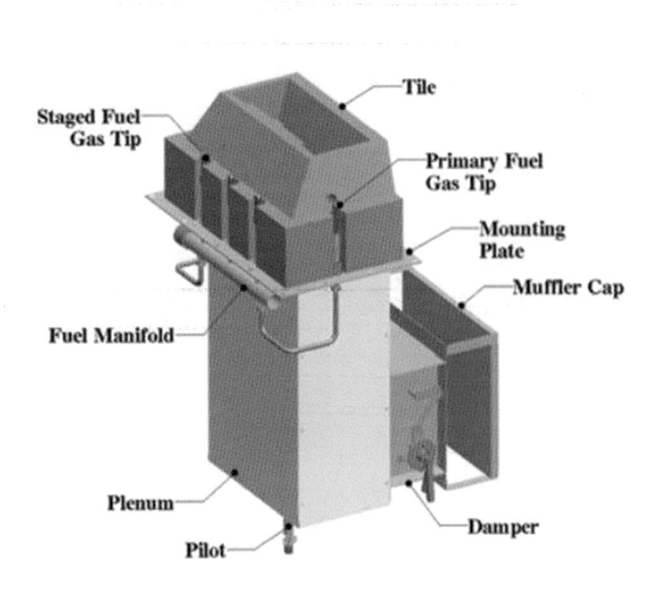

Figure 5.28 Conventional wall-fired burner. Source: C. E. Baukal, Jr. (ed.), *The John Zink Hamworthy Combustion Handbook, Volume 3: Applications*, CRC Press, Boca Raton, FL, 2013

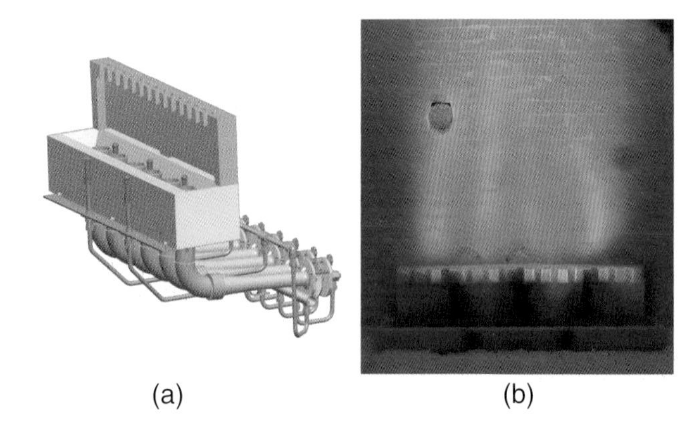

Figure 5.29 Ultra-low-NO_x wall-fired burner. Source: C. E. Baukal, Jr. (ed.), *The John Zink Hamworthy Combustion Handbook, Volume 3: Applications*, CRC Press, Boca Raton, FL, 2013

Figure 5.30 Rectangular ("flat") flame burners firing vertically upward along a wall in a fired heater. Source: C. E. Baukal, Jr. (ed.), *The John Zink Hamworthy Combustion Handbook, Volume 3: Applications*, CRC Press, Boca Raton, FL, 2013

5.2.1.3 *Wall-Fired Burners*

Michael Claxton, Tami Fischer, and Robert Luginbill

John Zink Hamworthy Combustion

Wall-fired burners are designed to fire along a wall so the wall can radiate heat to process tubes located fairly close to the burners. The burner outlet is commonly rectangular with a high length-to-width ratio to get as much flame coverage along the wall as possible. These burners are commonly called flat flame burners because of the shape. Figure 5.28 shows a conventional wall-fired burner commonly used in the floor of ethylene cracking furnaces. Figure 5.29 shows an ultra-low-NO_x version of this type of burner. Figure 5.30 shows flat flame burners firing up a wall on the left to heat process tubes on the right.

Keywords:
Wall-fired burners; ethylene cracking furnace

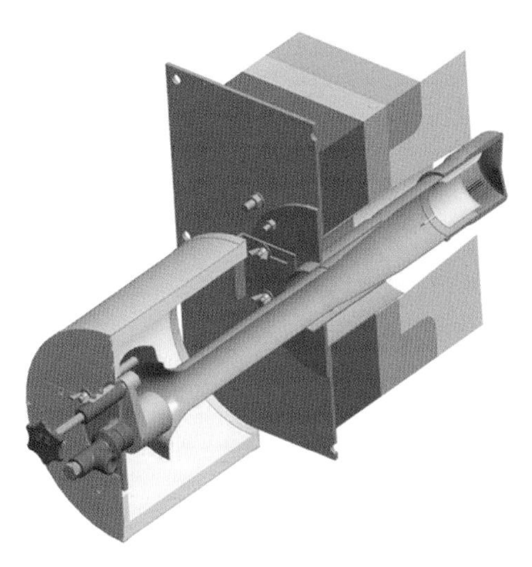

Figure 5.31 Natural draft radiant wall burner assembly. Source: C. E. Baukal, Jr. (ed.), *The John Zink Hamworthy Combustion Handbook*, Volume 3: Applications, CRC Press, Boca Raton, FL, 2013

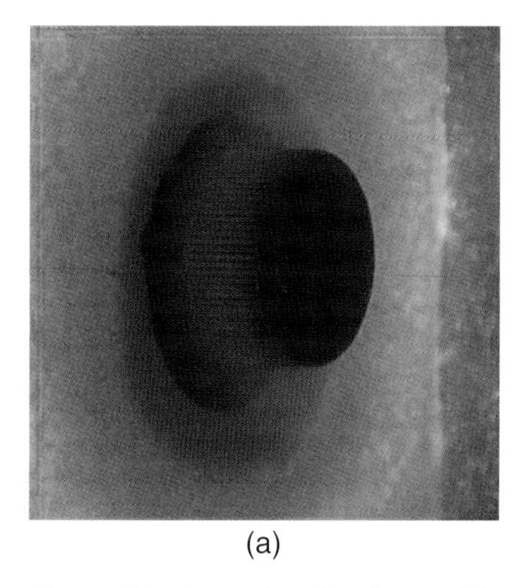

(a)

Figure 5.32a Natural draft radiant wall burner flame (side view). Source: C. E. Baukal, Jr. (ed.), *The John Zink Hamworthy Combustion Handbook*, Volume 3: Applications, CRC Press, Boca Raton, FL, 2013

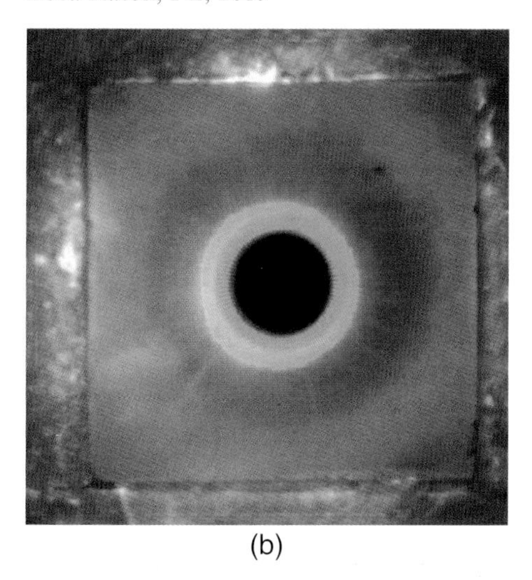

(b)

Figure 5.32b Natural draft radiant wall burner flame (front view). Source: C. E. Baukal, Jr. (ed.), *The John Zink Hamworthy Combustion Handbook*, Volume 3: Applications, CRC Press, Boca Raton, FL, 2013

Figure 5.33 Natural draft radiant wall burners in an ethylene furnace. Source: C. E. Baukal, Jr. (ed.), *The John Zink Hamworthy Combustion Handbook*, Volume 3: Applications, CRC Press, Boca Raton, FL, 2013

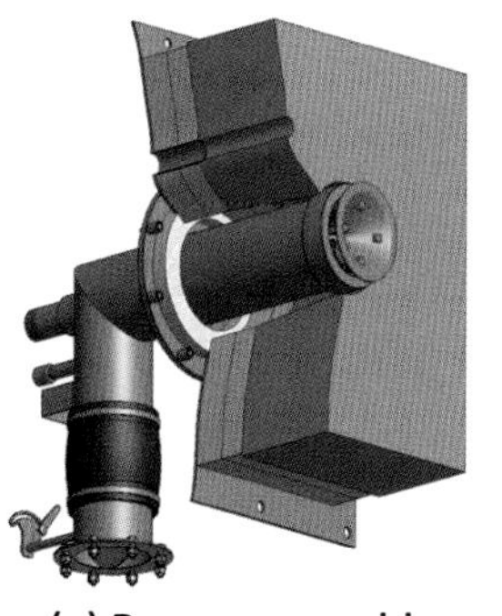

(a) Burner assembly

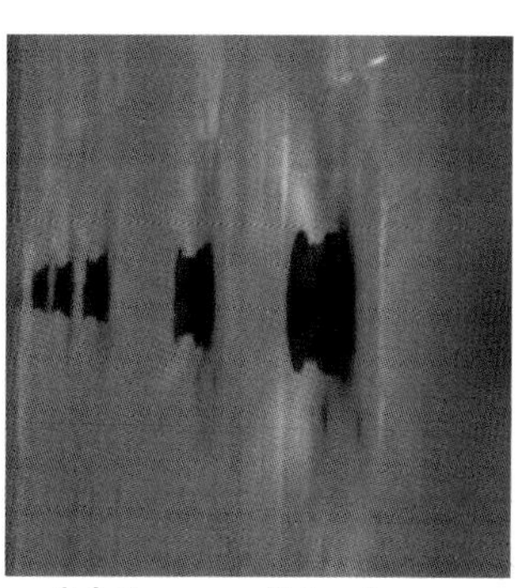

(b) Burners firing in an

Figure 5.34 Forced draft radiant wall burner. Source: John Zink Hamworthy Combustion (Tulsa, OK)

5.2.1.4 *Radiant Wall Burners*

Michael Claxton and Tami Fischer

John Zink Hamworthy Combustion

Figure 5.31 is a cutaway view of a natural draft radiant wall burner commonly used in ethylene cracking furnaces with a typical capacity of approximately 0.3 MW. Figures 5.32a and 5.32b show side and front views, respectively, of this burner firing in a test furnace. Figure 5.33 shows an array of these burners firing in an ethylene cracking furnace at approximately 1,200°C. Figure 5.34 shows a forced draft radiant wall burner: (a) assembly drawing; (b) burners firing in an ethylene furnace.

Keywords:

Radiant wall burners; ethylene cracking furnace

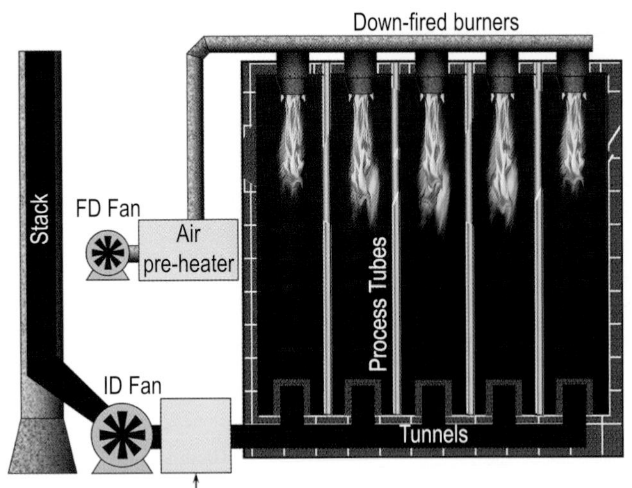

Figure 5.35 Downfired reformer schematic. Source: C. E. Baukal, Jr. (ed.), *The John Zink Hamworthy Combustion Handbook, Volume 1: Fundamentals and Volume 3: Applications*, CRC Press, Boca Raton, FL, 2013

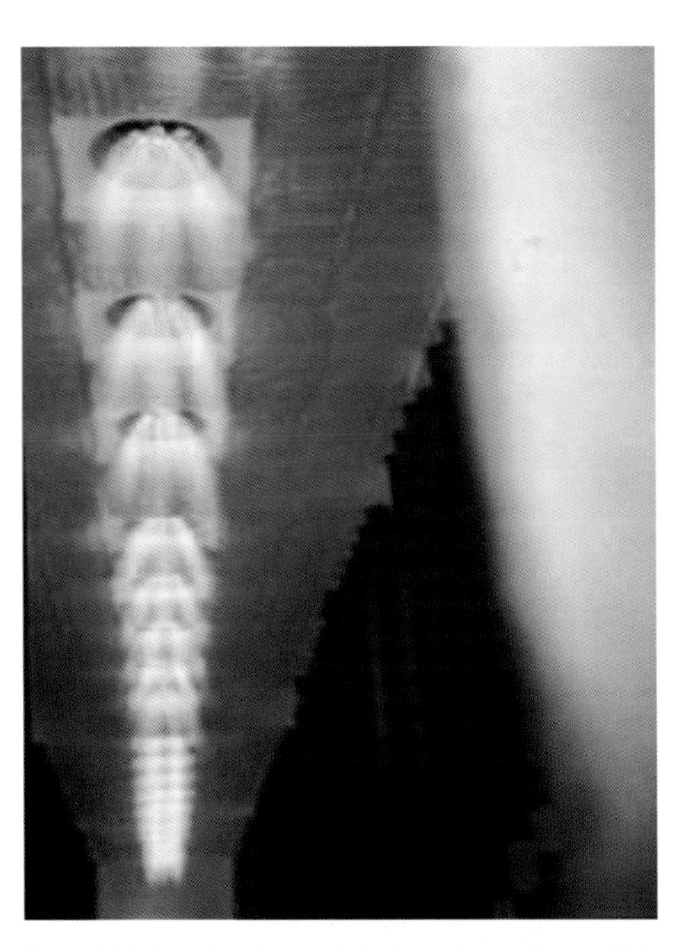

Figure 5.36 Downfired burners firing in a reformer. Source: C. E. Baukal, Jr. (ed.), *The John Zink Hamworthy Combustion Handbook, Volume 1: Fundamentals and Volume 3: Applications*, CRC Press, Boca Raton, FL, 2013

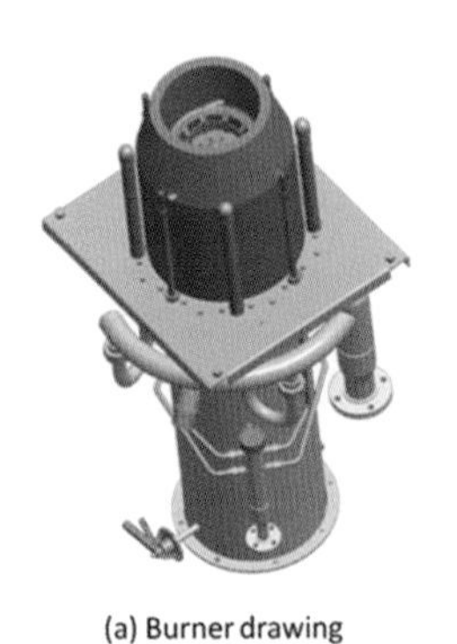

(a) Burner drawing

(b) Front view of the burner firing in a furnace

Figure 5.37 Example of a downfired burner. Source: C. E. Baukal, Jr. (ed.), *The John Zink Hamworthy Combustion Handbook, Volume 3: Applications*, CRC Press, Boca Raton, FL, 2013

Figure 5.38 Example of a downfired burner firing on naphta. Source: C. E. Baukal, Jr., *Industrial Combustion Pollution and Control*, Marcel Dekker, New York, 2004

5.2.1.5 *Downfired Burners*

Wes Bussman, Michael Claxton, Tami Fischer, Charles E. Baukal, Jr.

John Zink Hamworthy Combustion

Downfired burners are mounted in the roof of the furnace and fired downward as shown in Figures 5.35 and 5.36. They are commonly used in reforming furnaces used to produce hydrogen (H_2) and ammonia (NH_3). Figures 5.37a and 5.37b show a drawing of a typical burner and a closer view of a single burner firing, respectively. These burners often fire on two separate fuels simultaneously where the primary fuel is typically natural gas and the secondary fuel is an off-gas generated from the production of hydrogen. That secondary fuel gas typically has high concentrations of carbon monoxide (CO) and hydrogen (H_2) and is at a low pressure. Figure 5.38 shows a downfired burner firing on naphtha, which is a common liquid fuel used in this application.

Keywords:

Downfired burners; hydrogen reformers

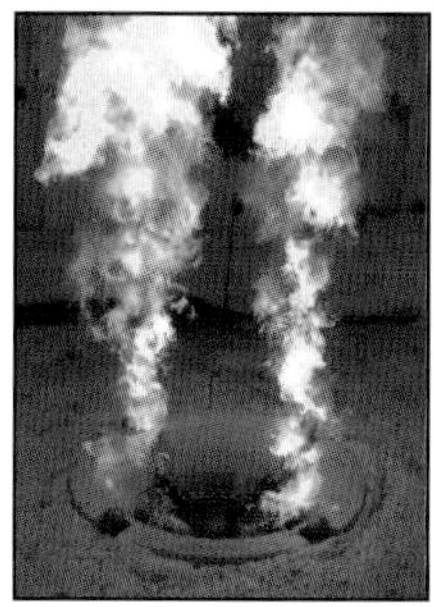

(a) 50% propylene / 50% natural gas at 3.7 psig (26 kPa).

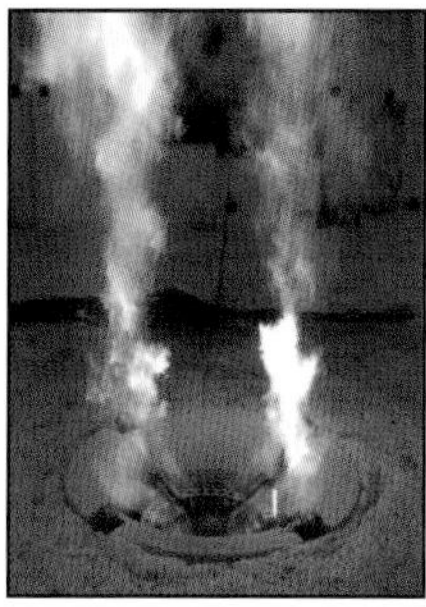

(b) 25% propylene / 75% natural gas at 4.2 psig (29 kPa).

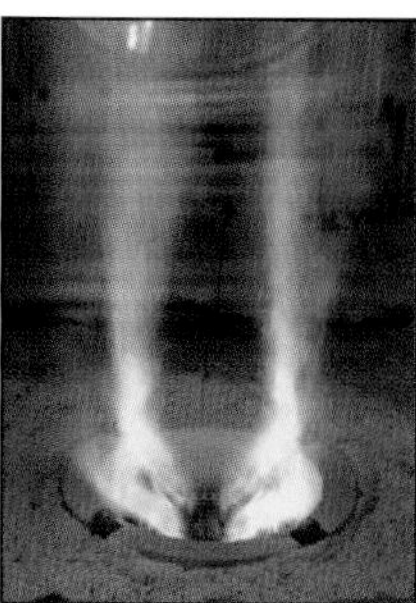

(c) 100% natural gas at 5.2 psig (36 kPa).

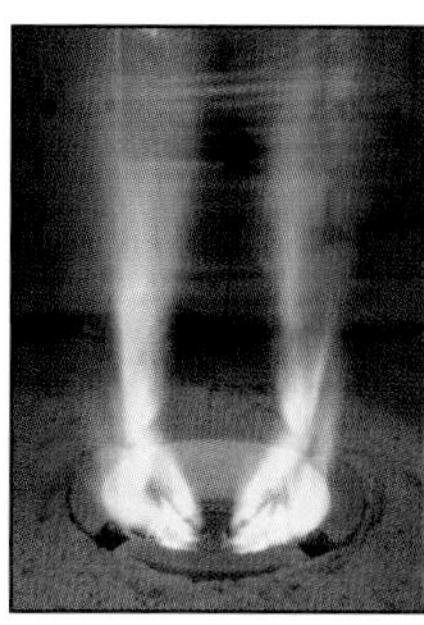

(d) 50% hydrogen / 50% natural gas at 7.1 psig (49 kPa).

Figure 5.39 Changing fuel composition while maintaining a constant heat release of 3.2×10^6 Btu/hr (0.94 MW), furnace temperature of 1,375°F (746°C), and excess O_2 of 3% dv. Source: John Zink Hamworthy Combustion (Tulsa, OK)

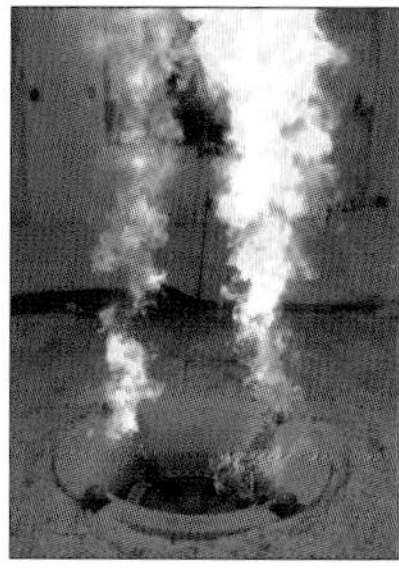

(a) 50% propylene / 50% natural gas, firing rate = 4.6×10^6 Btu/hr (1.3 MW), O2 = 3.0% dv

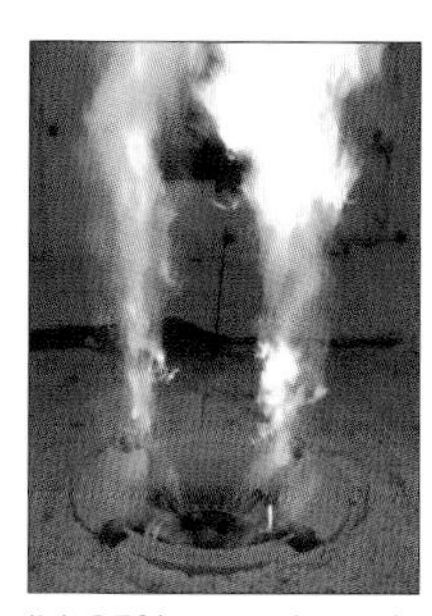

(b) 25% propylene / 75% natural gas, firing rate = 4.1×10^6 Btu/hr (1.2 MW), O2 = 4.6% dv

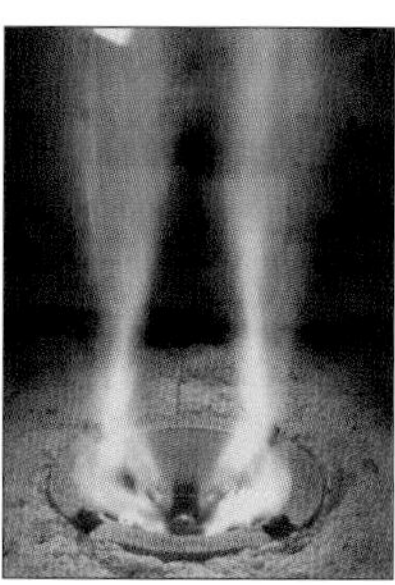

(c) 100% natural gas, firing rate = 3.7×10^6 Btu/hr (1.1 MW), O2 = 6.2% dv

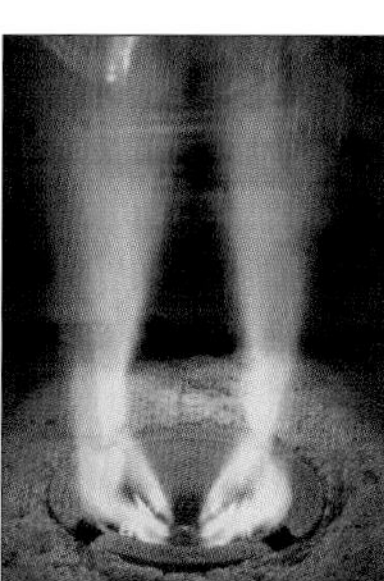

(d) 50% hydrogen / 50% natural gas, firing rate = 3.2×10^6 Btu/hr (0.94 MW), O2 = 9.1% dv

Figure 5.40 Changing fuel composition while maintaining a constant fuel pressure of 7 psig (48 kPa). Source: John Zink Hamworthy Combustion (Tulsa, OK)

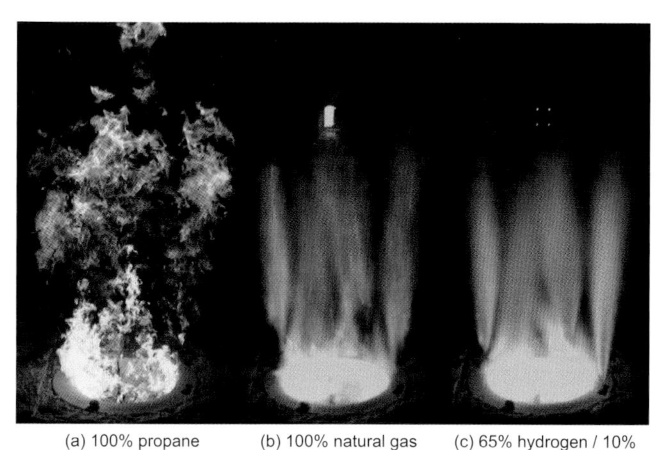

(a) 100% propane (b) 100% natural gas (c) 65% hydrogen / 10% propane / 25% natural gas

Figure 5.41 Flame shapes for different fuel compositions at a constant firing rate of 2.5×10^6 Btu/hr (0.73 MW). Source: John Zink Hamworthy Combustion (Tulsa, OK)

5.2.2 Burner Operations

5.2.2.1 *Gaseous Fuel Effects*

Wes Bussman

John Zink Hamworthy Combustion

The fuel composition has a large impact on the flame characteristics such as length, width, color, and intensity when the burner fuel nozzles have fixed openings. Figure 5.39 shows the effect of the fuel composition for a fixed firing rate where the fuel pressure is varied to maintain the firing rate. Figure 5.40 shows the same fuels used in Figure 5.39 where the fuel pressure remained constant while the firing rate and excess O_2 varied. Figure 5.41 shows three different fuels firing through the same burner at a constant firing rate. This shows the dramatic impact of fuel gas composition for fixed fuel orifices.

Keywords:
Gaseous fuel composition

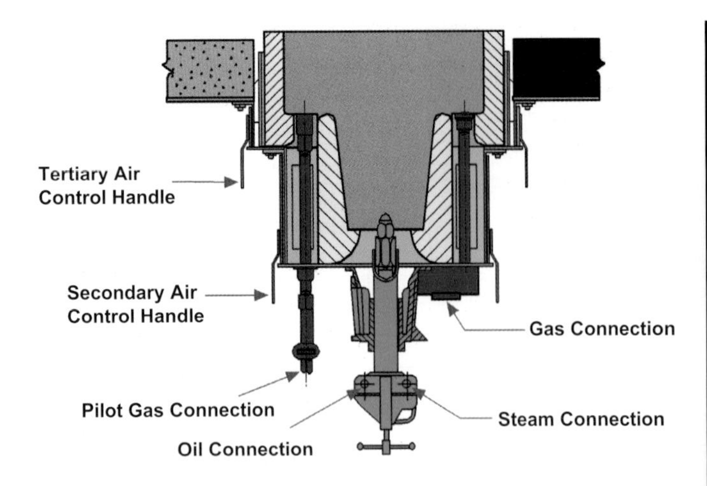

Figure 5.42 Combination burner cross-sectional drawing Source: C. E. Baukal, Jr. (ed.), *The John Zink Hamworthy Combustion Handbook, Volume 1: Fundamentals, Volume 2: Design and Operations, and Volume 3: Applications*, CRC Press, Boca Raton, FL, 2013

Figure 5.43 Oil atomization. Source: C. E. Baukal, Jr. (ed.), *The John Zink Hamworthy Combustion Handbook, Volume 1: Fundamentals, Volume 2: Design and Operations, and Volume 3: Applications*, CRC Press, Boca Raton, FL, 2013

Figure 5.44 Burner firing no. 6 oil. Source: C. E. Baukal, Jr. (ed.), *The John Zink Hamworthy Combustion Handbook, Volume 1: Fundamentals, Volume 2: Design and Operations, and Volume 3: Applications*, CRC Press, Boca Raton, FL, 2013

5.2.2.2 *Oil Firing*

I-Ping Chung, Michael Claxton, John Ackland, and Bill Johnson

John Zink Hamworthy Combustion

Liquid fuels are more challenging to fire because of the need to atomize the liquid into very fine droplets to get proper combustion. Figure 5.42 shows a drawing of a typical dual fuel burner capable of firing a gaseous or liquid fuel. Figure 5.43 shows the fine atomization required for proper combustion. Figure 5.44 is a burner firing heavy (no. 6) oil. Figure 5.45 shows a comparison of the same burner at approximately the same firing rate and furnace temperature firing on (a) oil and (b) refinery fuel gas. The oil flame is much more luminous than the gas flame.

Keyword:
Oil firing

Figure 5.45 Combination burner firing on (a) oil and (b) gas. Source: C. E. Baukal, Jr. (ed.), *The John Zink Hamworthy Combustion Handbook, Volume 1: Fundamentals, Volume 2: Design and Operations, and Volume 3: Applications*, CRC Press, Boca Raton, FL, 2013

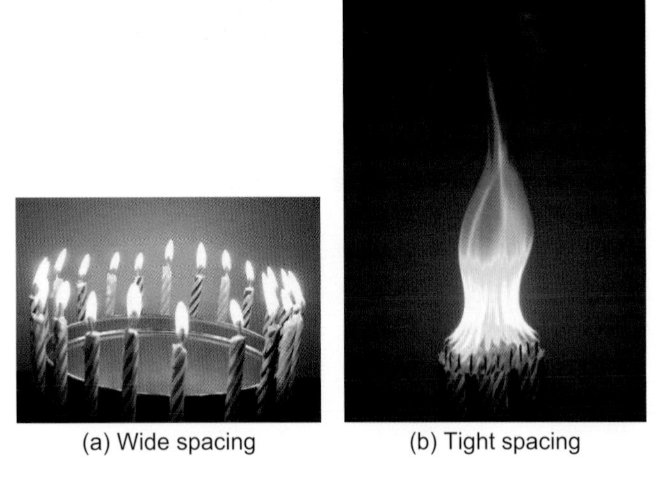

Figure 5.46 Birthday candles in a circle. Source: C. E. Baukal, Jr. (ed.), *The John Zink Hamworthy Combustion Handbook, Volume 1: Fundamentals*, CRC Press, Boca Raton, FL, 2013

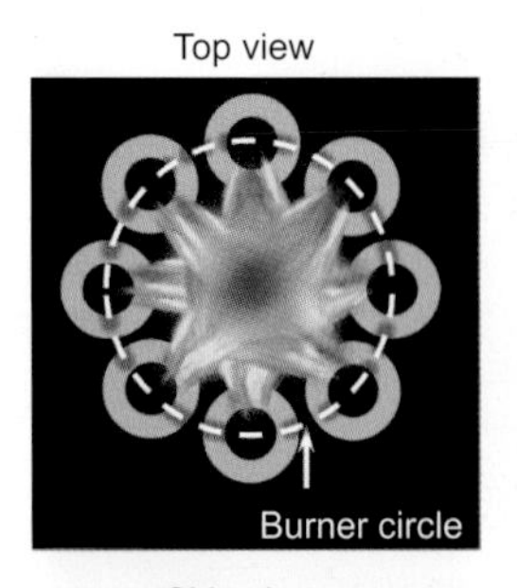

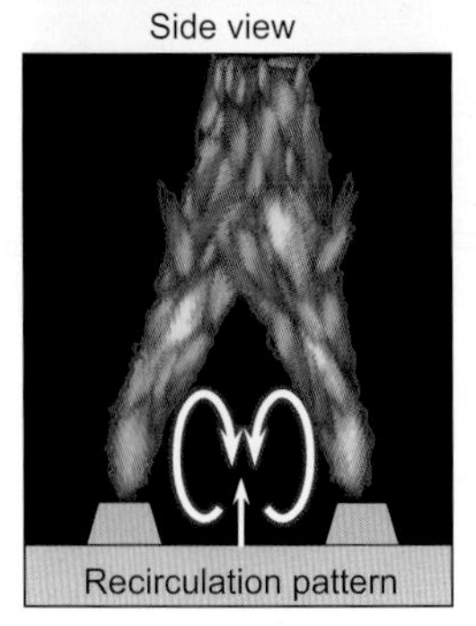

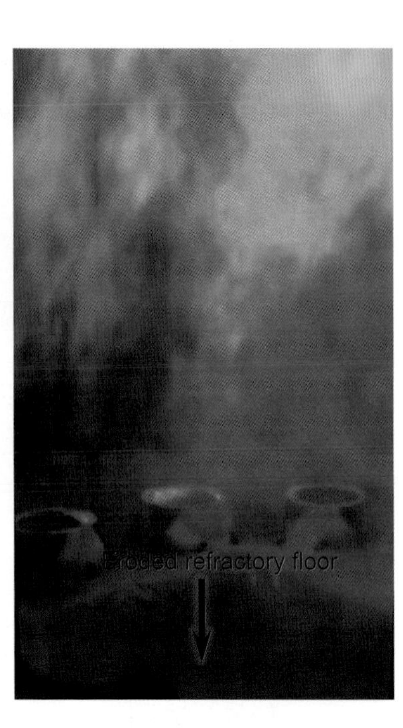

Figure 5.47 Vertical cylindrical furnace with burners too closely spaced in a circle. Source: C. E. Baukal, Jr. (ed.), *The John Zink Hamworthy Combustion Handbook, Volume 1: Fundamentals*, CRC Press, Boca Raton, FL, 2013

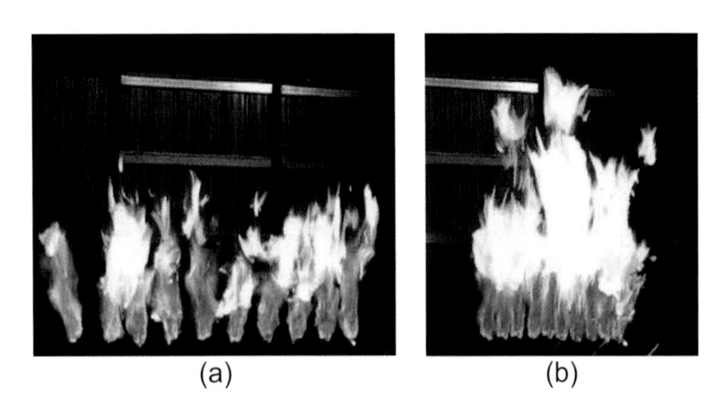

Figure 5.48 Diffusion burners arranged in a straight line: (a) burners spaced so no flame interaction and (b) burners at half the spacing shown in (a). Source: C. E. Baukal, Jr. (ed.), *The John Zink Hamworthy Combustion Handbook, Volume 1: Fundamentals*, CRC Press, Boca Raton, FL, 2013

5.2.2.3 *Burner Spacing*

Wes Bussman

John Zink Hamworthy Combustion

The spacing between burners is critical to performance in multi-burner systems. Figure 5.46 shows what happens when the spacing between birthday candles changes. Individual flames coalesce into one large flame when the spacing is too close. Figure 5.47 shows an actual example in a vertical cylindrical furnace where the burners are in a circle but too closely spaced, which does not allow for flue gas flow between the burners into the center. This creates a low-pressure region that actually caused erosion of the refractory floor. Figure 5.48 shows the effect of linear burner spacing on flame length and interaction. Burners are normally designed to have minimal interaction with adjacent burners. Excessive flame interaction can produce much larger flames than desired.

Keywords:
Burner spacing; flame-to-flame interaction

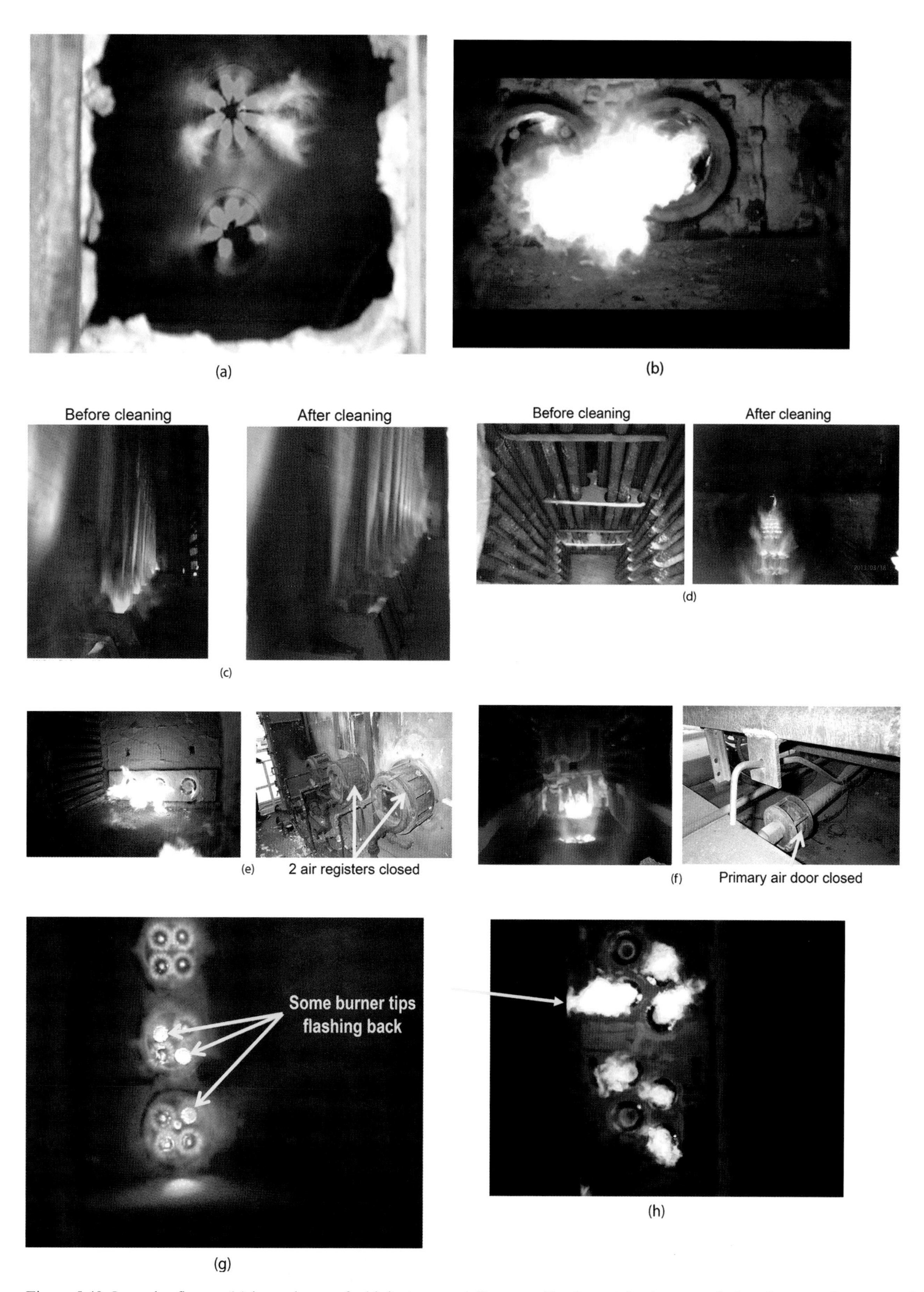

Figure 5.49 Irregular flames: (a) lower burner fuel injectors partially or totally plugged forcing more fuel to the upper burner causing yellow tails on some of the flames due to lack of air, (b) flames impinging on each other, (c) burners firing up a wall in an ethylene furnace before and after cleaning, (d) cabin heater before and after the burners were cleaned, (e) flames with 2 air registers closed, (f) rear flame with primary air door closed, (g) some burner tips flashing back, (h) flame impingement against the left side wall, and (i) and (j) flame impingement against the convection section tubes in a cabin heater. Source: C. E. Baukal, Jr. (ed.), *The John Zink Hamworthy Combustion Handbook, Volume 1: Fundamentals*, CRC Press, Boca Raton, FL, 2013

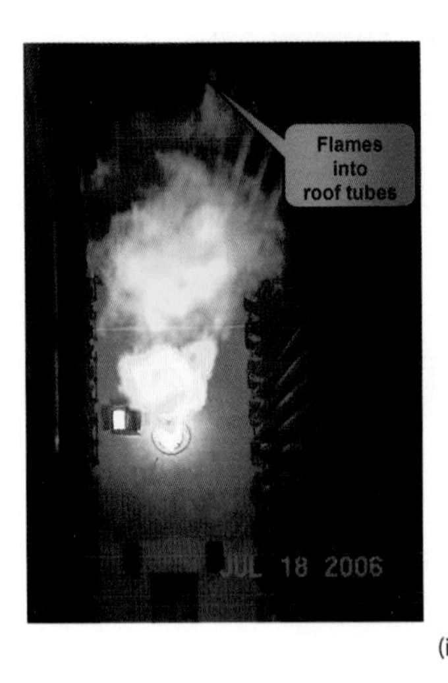

(i)

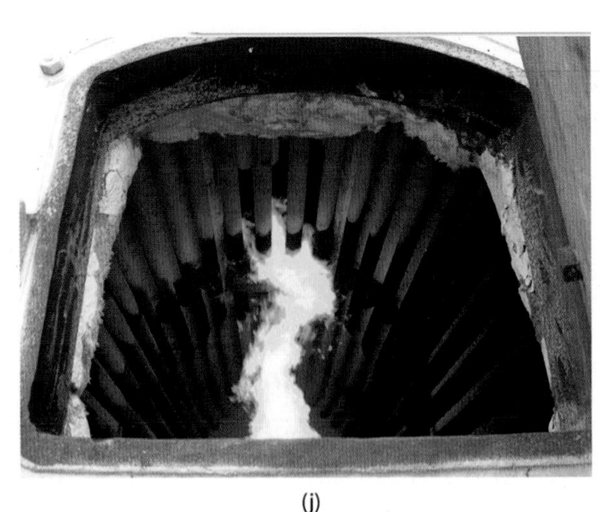

(j)

Figure 5.49 (*cont.*)

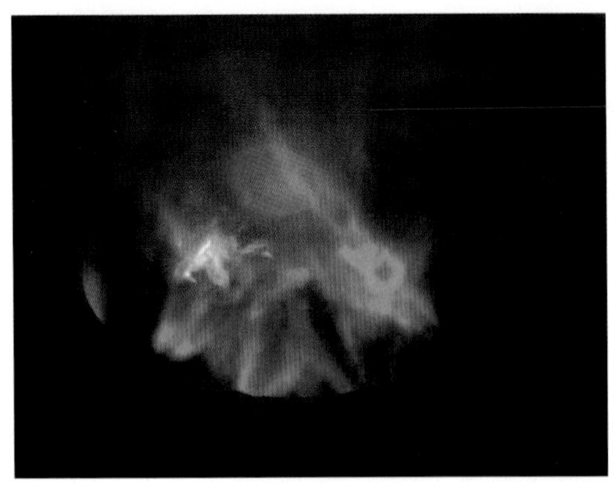

(a) With plugged tips
(before cleaning)

Figure 5.50a Floor-fired burner with (a) plugged tips (fuel injectors) and without plugged tips. Source: C. E. Baukal, Jr. (ed.), *The John Zink Hamworthy Combustion Handbook, Volume 1: Fundamentals*, CRC Press, Boca Raton, FL, 2013

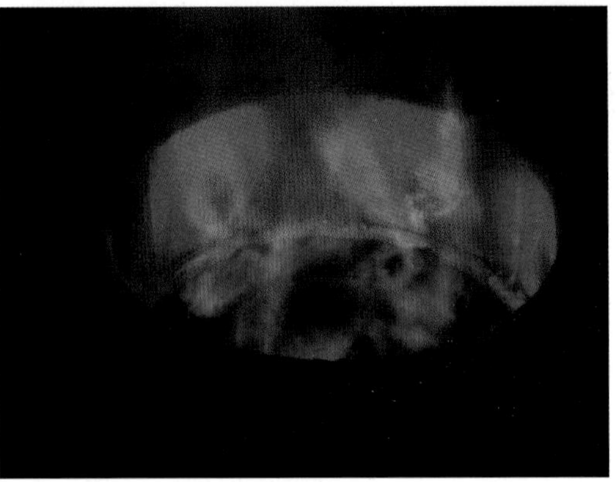

(b) Without plugged tips
(after cleaning)

Figure 5.50b Floor-fired burner after plugged tips (fuel injectors) have been cleaned. Source: C. E. Baukal, Jr. (ed.), *The John Zink Hamworthy Combustion Handbook, Volume 1: Fundamentals*, CRC Press, Boca Raton, FL, 2013

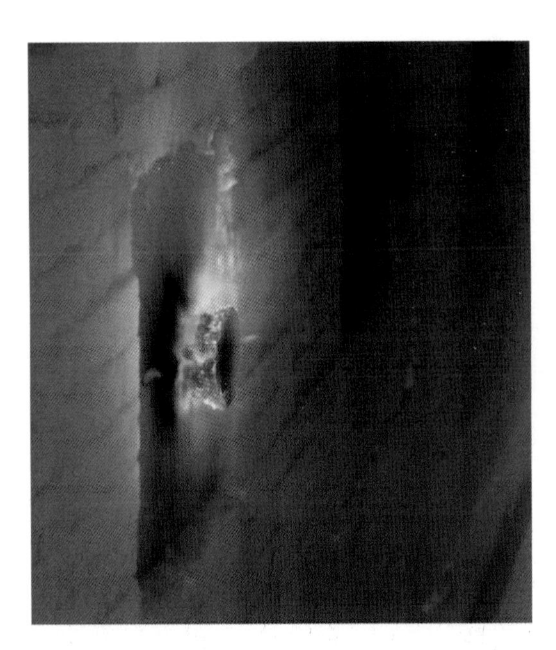

Figure 5.51 Flame flashing back inside a radiant wall burner. Source: C. E. Baukal, Jr. (ed.), *The John Zink Hamworthy Combustion Handbook, Volume 1: Fundamentals*, CRC Press, Boca Raton, FL, 2013

5.2.2.4 Examples of Burner Problems

Wes Bussman and Bill Johnson
John Zink Hamworthy Combustion

There are numerous possible burner problems, so only a few examples are given here. Figure 5.49 shows two wall-mounted burners where some of the fuel injectors on the bottom burner are plugged (no flames in parts of the lower half). Figure 5.50a shows a burner with some plugged fuel injectors before cleaning. Figure 5.50b shows the same burner after the injectors have been cleaned. Figure 5.51 shows a flame flashing back inside a single radiant wall burner. The likely cause is some type of pluggage. Note the other burners near it are not flashing back.

Keywords:
Burner problems; pluggage; flashback

Figure 5.52 Adjusting draft and excess O_2 for flat flame burners fired horizontally along the floor of a cabin heater (a) before and (b) after adjustment. Source: John Zink Hamworthy Combustion (Tulsa, OK)

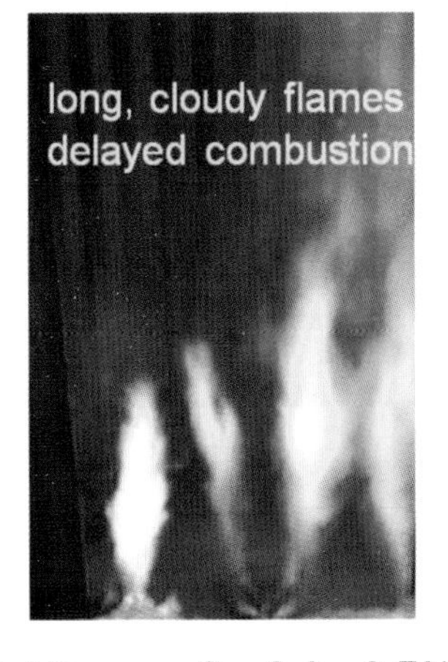

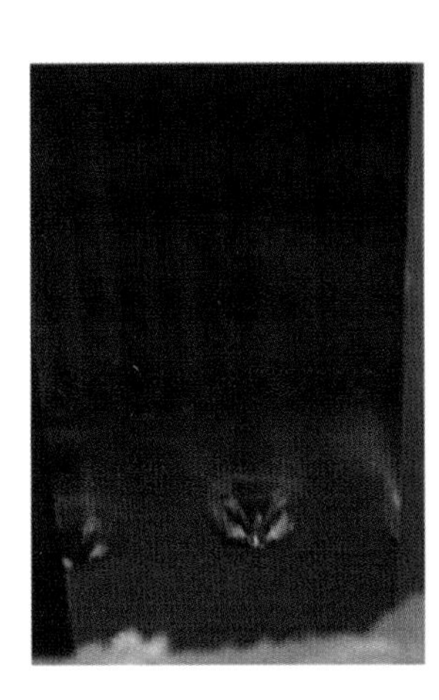

Figure 5.53 Round flame burners firing vertically upward in a vertical cylindrical heater: (a) before and (b) after adjustment. Source: John Zink Hamworthy Combustion (Tulsa, OK)

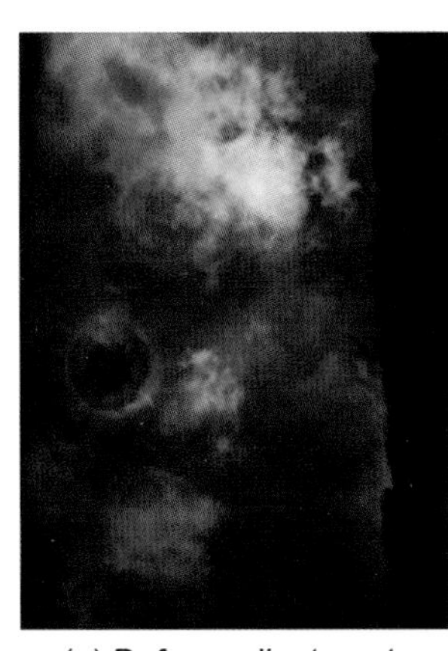

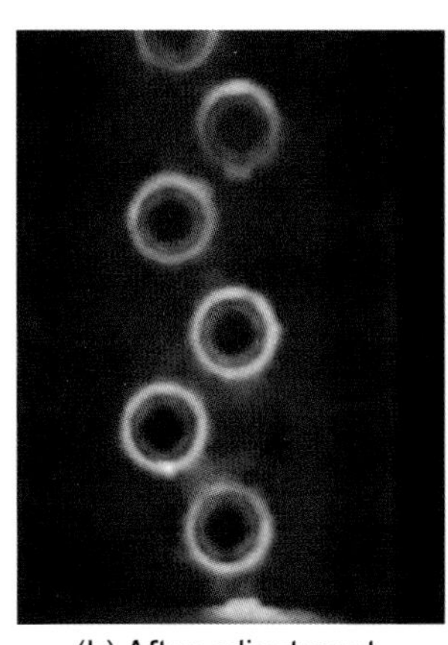

Figure 5.54 Adjusting draft and excess O_2 for round burners fired from an end wall of a cabin heater: (a) before and (b) after adjustment. Source: John Zink Hamworthy Combustion (Tulsa, OK)

5.2.2.5 *Adjusting Burners*

Mike Pappe and Robert Luginbill
John Zink Hamworthy Combustion

These photos show the improper operating conditions before adjusting the burners and the proper operating conditions after adjusting the burners. Figure 5.52 shows flat flame burners firing horizontally across the floor of a cabin heater before and after adjusting the draft and excess O_2. Figure 5.53 shows round burners firing vertically upward in a vertical cylindrical heater before and after adjusting the excess O_2. Figure 5.54 shows round flame burners firing horizontally from the end wall of a cabin heater before and after adjusting the draft and excess O_2.

Keywords:
Adjusting burners; draft; excess oxygen

Figure 5.55 Heater with original floor-fired burners before replacement with ultra-low-NO_x burners. Source: C. E. Baukal, Jr., *Industrial Combustion Pollution and Control*, Marcel Dekker, New York, 2004

Figure 5.56 Heater after replacement with ultra-low-NO_x burners. Source: C. E. Baukal, Jr., *Industrial Combustion Pollution and Control*, Marcel Dekker, New York, 2004

5.2.2.6 Replacing Burners

Jim Seebold

Chevron

Richard Waibel

John Zink Hamworthy Combustion

Figure 5.55 shows conventional burners firing refinery fuel gas through tunnels horizontally across the floor of a cabin heater. A limiting factor for the firing rate was the maximum refractory temperature along the floor. NO_x was approximately 180 ppmvd at 3% excess O_2. Figure 5.56 shows a similar furnace after the conventional burners were replaced with ultra-low-NO_x burners. Notice the much more uniform refractory temperature, which permitted increasing the firing rate for more production. The NO_x was reduced to approximately 20 ppmvd for almost a 90% reduction.

Reference

1. J. G. Seebold, R. G. Miller, G. W. Spesert, D. E. Beckley, D. J. Coutu, R. T. Waibel, et al., Developing and retrofitting ultra low NO_x burners in a refinery furnace. Proceedings of Joint International Combustion Symposium, AFRC/JFRC/IEA 2001, Section 2C, Paper 2, Kauai, Hawaii, September 2001.

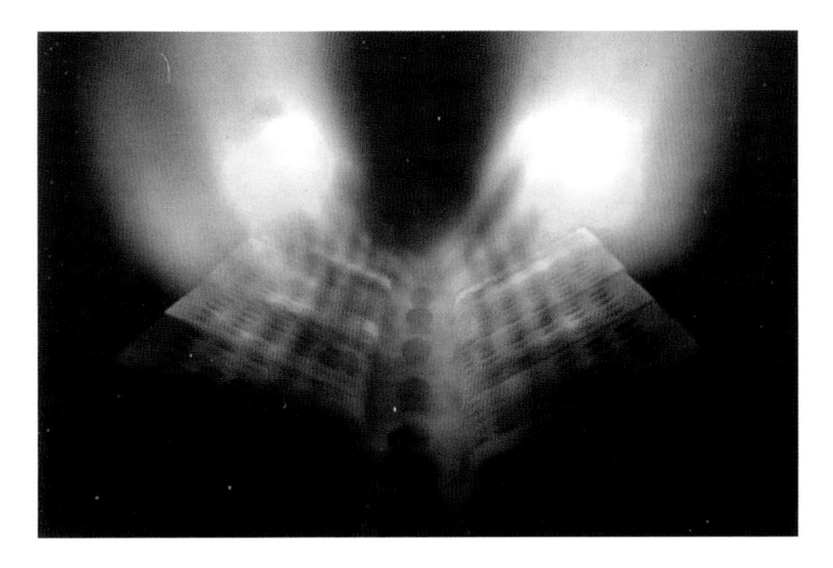

Figure 5.57 Duct burner firing on natural gas. Source: Peter F. Barry, Stephen L. Somers, and Stephen B. Londerville, "Duct Burners," in *Coen Hamworthy Combustion Handbook*, edited by S. B. Londerville and C. E. Baukal, CRC Press, Boca Raton, FL, 2013

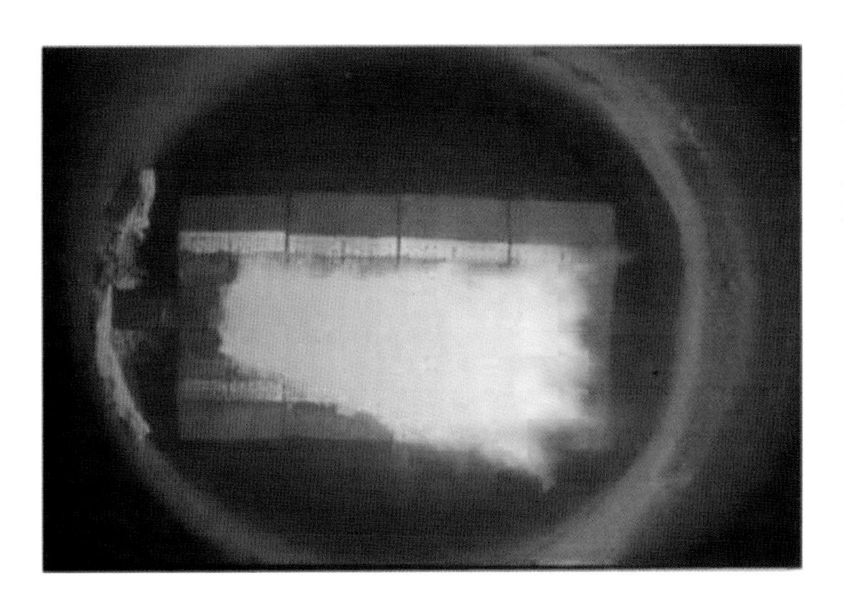

Figure 5.58 Duct burners firing on oil. Source: Peter F. Barry, Stephen L. Somers, and Stephen B. Londerville, "Duct Burners," in *Coen Hamworthy Combustion Handbook*, edited by S. B. Londerville and C. E. Baukal, CRC Press, Boca Raton, FL, 2013

5.3 Power Generation

5.3.1 Duct Burners

Stephen B. Londerville

Coen

Duct burners are commonly used in the outlet of stationary gas turbines used for power generation to increase the production for a given size system. The combustion "air" actually consists of the exhaust products from the turbine, which typically contains 11–16% oxygen by volume and is at an elevated temperature of 315–540°C (600–1,000°F). Figure 5.57 shows a grid-style duct burner firing on natural gas. Figure 5.58 shows a side-fired oil burner firing across a grid.

Keyword:
Duct burners

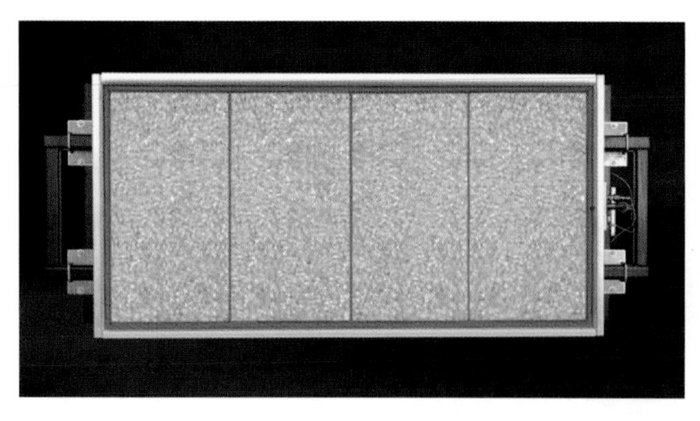

Figure 5.59 Porous refractory burner firing on natural gas. Source: Thomas M. Smith and C. E. Baukal, Jr., "Thermal Radiation Burners," in C. E. Baukal, Jr. (ed.), *Industrial Burners Handbook*, CRC Press, Boca Raton, FL, 2004

Figure 5.61 Schematic of the drying section of a Fourdrinier paper-making machine using porous refractory burners. Source: Thomas M. Smith and C. E. Baukal, Jr., "Thermal Radiation Burners," in C. E. Baukal, Jr. (ed.), *Industrial Burners Handbook*, CRC Press, Boca Raton, FL, 2004

Figure 5.60 Porous refractory burner used to dry textiles. Source: Thomas M. Smith and C. E. Baukal, Jr., "Thermal Radiation Burners," in C. E. Baukal, Jr. (ed.), *Industrial Burners Handbook*, CRC Press, Boca Raton, FL, 2004

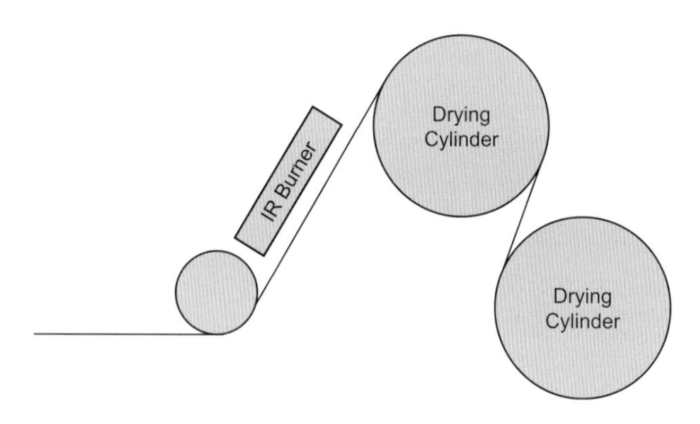

Figure 5.62 Porous refractory burners used to paper. Source: Thomas M. Smith and C. E. Baukal, Jr., "Thermal Radiation Burners," in C. E. Baukal, Jr. (ed.), *Industrial Burners Handbook*, CRC Press, Boca Raton, FL, 2004

5.4 Infrared Heating and Drying

5.4.1 Thermal Radiation Burner

Gerard J. Lucidi
Marsden Inc.

Figure 5.59 shows the firing side of a Marsden Natural Gas Instant On/Off infrared (IR) burner used in a wide range of applications, some of which are included here. The burners operate through a full range of temperature from 1,100°F to 1,850°F (590–1,010°C). A common application is to dry moving webs (sheets) of paper or textiles as shown in Figure 5.60. Figure 5.61 is a schematic of the drying section of a Fourdrinier paper-making machine. The paper is made from a wet pulp slurry that is dried by traveling over steam-heated drums. However, paper is a good insulator, so the drier the paper is, the poorer the thermal conduction and the poorer the thermal efficiency of the steam-heated cylinders. IR burners are effective for enhancing drying near the end of the process because they do not rely on conduction, but rather on penetrating thermal radiation. Figure 5.62 shows an example of that application. A common problem for a Fourdrinier is that it

Figure 5.63 Porous refractory burners used in moisture profiling where the firing rate of each segment is determined by upstream moisture sensors. Source: Thomas M. Smith and C. E. Baukal, Jr., "Thermal Radiation Burners," in C. E. Baukal, Jr. (ed.), *Industrial Burners Handbook*, CRC Press, Boca Raton, FL, 2004

Figure 5.64 Porous refractory burners used to preheat and initiate the evaporation process following the application of water based coatings. Source: Thomas M. Smith and C. E. Baukal, Jr., "Thermal Radiation Burners," in C. E. Baukal, Jr. (ed.), *Industrial Burners Handbook*, CRC Press, Boca Raton, FL, 2004

often produces an uneven moisture profile across the width of the paper as determined by optical sensors viewing the high-speed moving paper web. Paper is typically dried to some maximum moisture level. The problem is that some segments of the paper web may be dried to significantly lower moisture levels because of the uneven profile. These drier sections have reduced quality and can cause problems, for example, in copiers and printers. Additionally, paper is sold in part by weight, so removing more water than needed reduces revenue. Figure 5.63 shows an example of a moisture-profiling burner where the firing rate of each segment is determined by upstream moisture sensors. These burners have extremely fast reaction times and can adjust quickly to moisture changes. IR burners can be used to dry coatings on paper. Figure 5.64 shows a 31.5″ × 350″ (80 cm × 890 cm) IR Drying Frame utilizing an emitter to preheat and initiate the evaporation process following the application of water-based coatings. Each Drying Frame transfers a maximum of 3 MMBtu/hr (0.9 MW). The emitters heat up to full power (>1,850°F or 1,010°C) in 5 s and cool to the touch in 1 s to dry the applied coating on lightweight paper. Figure 5.65 shows 11″ × 66″ (28 cm × 170 cm) Marsden IR burners for annealing glass tubes. Each burner has a maximum heat transfer of 217,800 Btu/hr (64 kW).

Figure 5.65 Porous refractory burners used to anneal glass tubes. Source: Thomas M. Smith and C. E. Baukal, Jr., "Thermal Radiation Burners," in C. E. Baukal, Jr. (ed.), *Industrial Burners Handbook*, CRC Press, Boca Raton, FL, 2004

Keywords:

Thermal radiation burner; infrared drying; glass; annealing; coated paper; moisture profiling

Figure 5.66 Steam-assisted flare with the steam off. Source: Robert Schwartz, Jeff White, and Wes Bussman, "Flares," in *The John Zink Combustion Handbook*, CRC Press, Boca Raton, FL, 2001

Figure 5.67 Steam-assisted flare with the steam just starting. Source: Robert Schwartz, Jeff White, and Wes Bussman, "Flares," in *The John Zink Combustion Handbook*, CRC Press, Boca Raton, FL, 2001

Figure 5.68 Steam-assisted flare with the steam at full rate. Source: Robert Schwartz, Jeff White, and Wes Bussman, "Flares," in *The John Zink Combustion Handbook*, CRC Press, Boca Raton, FL, 2001

5.5 Flares

5.5.1 Steam-Assisted Flares

Robert Schwartz, Jeff White, and Wes Bussman

John Zink Hamworthy Combustion

Steam is often used to entrain additional combustion air into flare flames to increase their smokeless capacity when the waste gas is at a relatively low pressure. Figure 5.66 shows a smoking flare before the assist steam has been turned on. Figure 5.67 shows the same flare just after the steam assist has been turned on. Figure 5.68 shows the steam assist at full flow where the smoking has been eliminated.

Keyword:
Air-assisted flares

Figure 5.69 Air-assisted flare with the blower off. Source: Robert Schwartz, Jeff White, and Wes Bussman, "Flares," in *The John Zink Combustion Handbook*, CRC Press, Boca Raton, FL, 2001

Figure 5.70 Air-assisted flare with the blower just starting. Source: Robert Schwartz, Jeff White, and Wes Bussman, "Flares," in *The John Zink Combustion Handbook*, CRC Press, Boca Raton, FL, 2001

Figure 5.71 Air-assisted flare with the blower speed increasing. Source: Robert Schwartz, Jeff White, and Wes Bussman, "Flares," in *The John Zink Combustion Handbook*, CRC Press, Boca Raton, FL, 2001

5.5.2 Air-Assisted Flares

Robert Schwartz, Jeff White, and Wes Bussman

John Zink Hamworthy Combustion

Blowers are sometimes used to supply additional combustion air to flares to increase their smokeless capacity when the waste gas is at a relatively low pressure. Figure 5.69 shows a smoking flare before the assist air has been turned on. Figure 5.70 shows the same flare just after the assist blower has been turned on. Figure 5.71 shows the air assist blower increasing its output. Figure 5.72 shows the assist blower at full capacity where the smoking has been eliminated.

Figure 5.72 Air-assisted flare with the blower at full rate. Source: Robert Schwartz, Jeff White, and Wes Bussman, "Flares," in *The John Zink Combustion Handbook*, CRC Press, Boca Raton, FL, 2001

Keyword:

Air-assisted flares

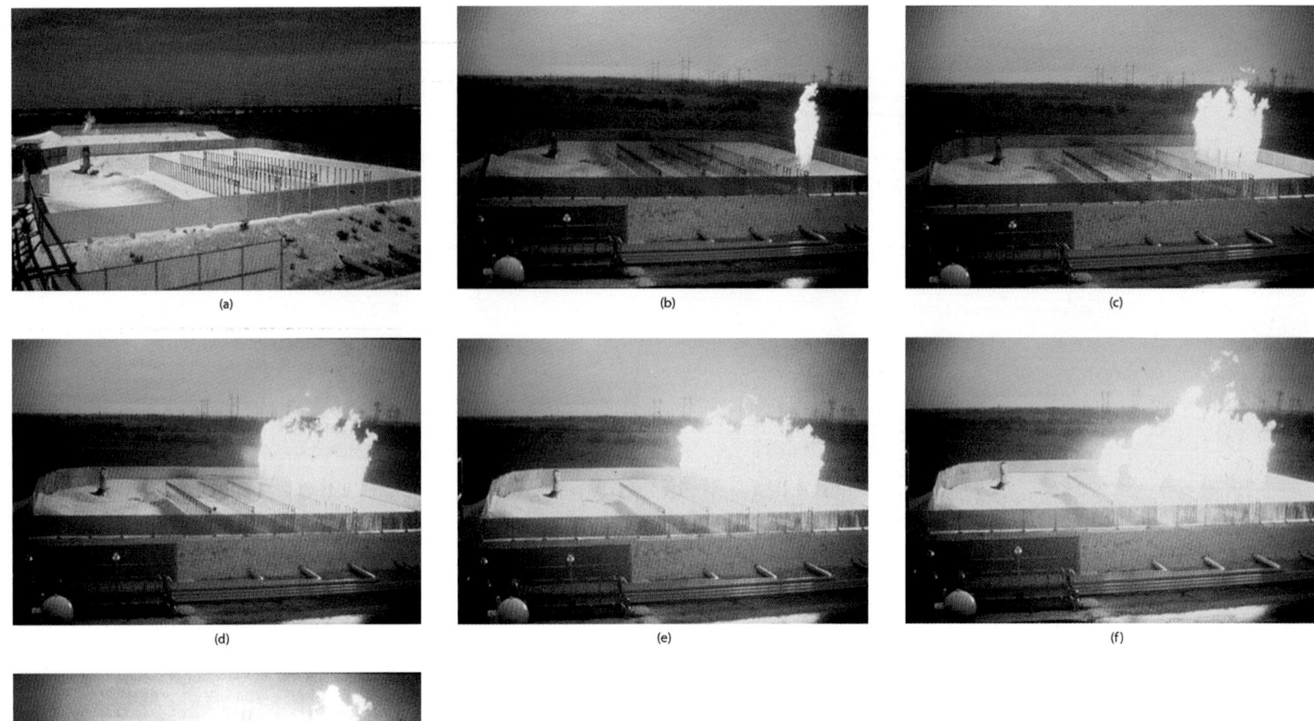

Figure 5.73 Ground flare with (a) no burners on increasing until (g) all burner stages are on. Source: Robert Schwartz, Jeff White, and Wes Bussman, "Flares," in *The John Zink Combustion Handbook*, CRC Press, Boca Raton, FL, 2001

Figure 5.74 Ground flare burner spacing test. Source: Robert Schwartz, Jeff White, and Wes Bussman, "Flares," in *The John Zink Combustion Handbook*, CRC Press, Boca Raton, FL, 2001

5.5.3 Ground Flares

Robert Schwartz, Jeff White, and Wes Bussman

John Zink Hamworthy Combustion

If waste gases are at a high enough pressure, then no assist media such as steam or assist air are needed, so unassisted burners can be used. However, two potential challenges are the very wide flow range that may be encountered and the need to conceal the flames from the neighbors around the plant. These are often accomplished by using many smaller burners in an array surrounded by a fence to substantially hide the flames. Figure 5.73 shows such an array with four rows of burners where more rows of burners come on as the waste gas pressure increases. This permits the burners to fire more optimally at their design capacity. It is desirable to make the array as small as possible to minimize the plot space. Figure 5.74 shows a test of ground flare burners in a single row to determine optimal spacing. If the burners are too close together, the flames will coalesce and become much longer, which is not desirable. If they are too far apart, then the array becomes larger and more costly.

Keyword:
Air-assisted flares

Figure 5.75 Flare radiation on an offshore oil rig. Source: John Zink Hamworthy Combustion (Tulsa, OK)

Figure 5.76 Flare without water injection. Source: John Zink Hamworthy Combustion (Tulsa, OK)

Figure 5.77 Flare with water injection. Source: John Zink Hamworthy Combustion (Tulsa, OK)

5.5.4 Flare Radiation Reduction

Kevin Leary and Charles E. Baukal, Jr.
John Zink Hamworthy Combustion

One of the reasons some land-based flares have flames so high above the ground is to minimize the thermal radiation at ground level that could injure people and damage equipment. In some applications it is not feasible to build a tall flare, such as on an offshore oil rig. Figure 5.75 shows one method for minimizing the flare radiation to the platform by having a boom that directs the flame away from the platform. Another method is to inject seawater into the flame where the water reduces the flame temperature and radiation. Figure 5.76 shows such a flare with the water off. Figure 5.77 shows the same flame after the water has been turned on. Notice the significant reduction in flame luminosity with water injection.

(a)

(b)

(c)

Figure 5.78 Flaring event: (a) start, (b) increasing, and (c) at a high rate. Source: Robert Schwartz, Jeff White, and Wes Bussman, "Flares," in *The John Zink Combustion Handbook*, CRC Press, Boca Raton, FL, 2001

5.5.5 Flare Problems

Robert Schwartz, Jeff White, Wes Bussman, and Zachary Kodesh

John Zink Hamworthy Combustion

Figure 5.78 shows an example of what can happen when liquid hydrocarbons are not removed from waste products going to a flare designed to handle only gaseous wastes. Much of the liquid falls out and burns on its way to the ground, which is often referred to as flaming rain. Figure 5.79 shows a high-wind condition that can cause a flare flame to pull down on the side of the flare tip, which can lead to equipment damage if this condition is sustained for too long.

Keywords:
Flare problems; flaming rain; flame pulldown

(a) High crosswind bends flame from vertical to nearly horizontal

(b) flame pull-down on flare stack

Figure 5.79 Flame pulldown due to high winds. Source: Wes Bussman, Zachary L. Kodesh, and Robert E. Schwartz, "Fundamentals of Fluid Dynamics," in *The John Zink Hamworthy Combustion Handbook, Volume 1: Fundamentals*, CRC Press, Boca Raton, FL, 2013

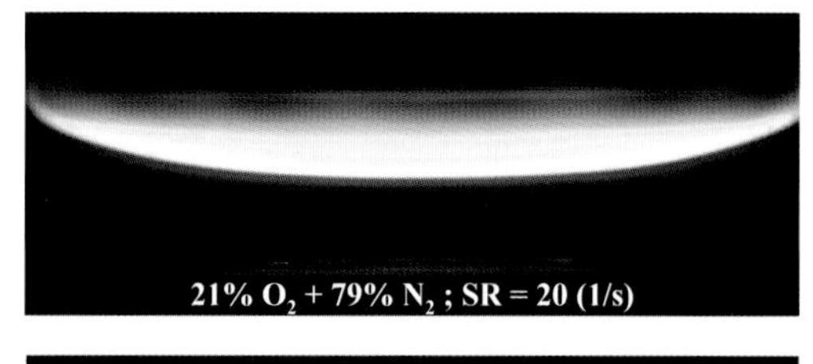

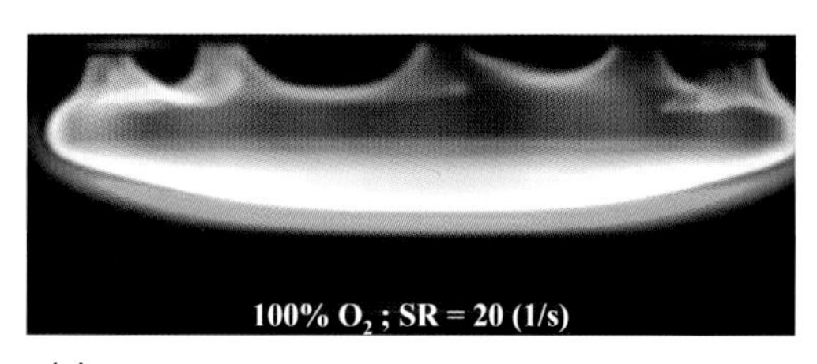

(a)

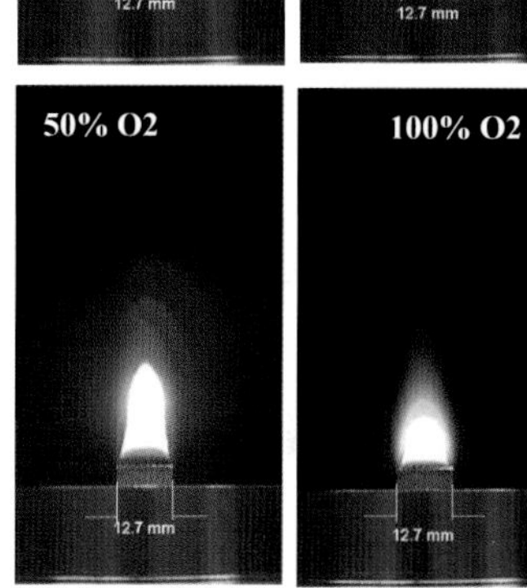

(b)

Figure 5.80 (a) Representative photographs of counterflow methane flames formed at various oxygen contents (21–100% O_2); (b) display methane co-flow flames formed by varying the oxygen content in the oxidizer from 21% to 100% O_2. Both flames are composed of two flame zones – a yellow zone and a blue zone. Source: Wilson Merchan-Merchan, Sergio Sanmiguel, Alexi Saveliev, and Stephen McCullam, "Soot Formation in Oxygen-Enhanced Combustion," Chapter 15 in *Oxygen-Enhanced Combustion*, 2nd edition, edited by C. E. Baukal, Jr., CRC Press, Boca Raton, FL, 2013

Figure 5.81 Oxy/fuel flames used in glass polishing. Source: Hisashi Kobayashi, Rémi Tsiava, and C. E. Baukal, Jr., "Burner Design," Chapter 18 in *Oxygen-Enhanced Combustion*, 2nd edition, edited by C. E. Baukal, Jr., CRC Press, Boca Raton, FL, 2013

(a)

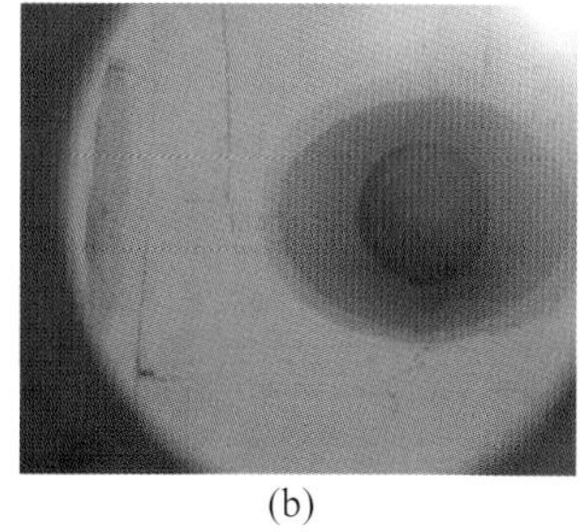

(b)

Figure 5.82 Flameless oxy/fuel technology. Source: Wlodzimierz Blasiak, Weihong Yang, Thomas Ekman, and Joachim von Schele, "Flameless Oxyfuel Combustion and Its Applications," Chapter 20 in Oxygen-Enhanced Combustion, 2nd edition, edited by C. E. Baukal, Jr., CRC Press, Boca Raton, FL, 2013

5.6 Oxygen-Enhanced Flames

Walmy Cuello Jimenez and Wilson Merchan-Merchan

University of Oklahoma

Sergei Zelepouga and Alexei Saveliev

North Carolina State University

Hisashi Kobayashi

Praxair

Rémi Tsiava

Air Liquide

Wlodzimierz Blasiak

Royal Institute of Technology

For the vast majority of industrial combustion applications, the oxidizer is air (approximately 21% O_2 by volume) because of its low cost and availability. However, there are applications where the oxygen content in the oxidizer is higher than 21%, which is often referred to as oxygen-enhanced combustion. The oxygen content can range from 21% all the way to 100% (typically referred to as oxy-fuel. Figure 5.80 shows how the flame structure can vary with the oxygen content in the oxidizer. Figure 5.81 shows an example of oxy-fuel flames used to polish glass. Figure 5.82 shows flameless oxy-fuel technology.

Keyword:

Oxygen-enhanced combustion

6 Fires

Edited by Michael J. Gollner

Introduction

Michael J. Gollner

Scientists have long studied fire in an effort to both understand the world around them and to prevent the destruction and devastation that uncontrolled fires can cause. Despite many advances in the understanding of fire phenomena, society offers continued challenges that require new approaches for the prevention and mitigation of unwanted fires. In this chapter, fire research is presented through a series of photographs that scale from small, buoyant flames in the laboratory up to large, uncontrolled wildfires and even fire whirls.

Our journey begins with a study of the most canonical problem in fire science, the pool fire, where liquid fuel in a pan is burned in a quiescent atmosphere. Interesting effects such as puffing and flickering are explored, as well as the influence of different fuels and configurations. Simulations of a pool fire show the complex instabilities that govern its behavior. Similar behavior is shown in very large fires, here up to 120 m in diameter. Another canonical problem in fire, upward flame spread, is presented in comparison to other drivers of the flame-spread process: orientation, ambient wind, and configuration. Together, the heat-release rate from a fire and flame-spread rate form the primary descriptors of material flammability, signifying the safety and appropriateness of materials for fire-safe design.

When design fails and accidents occur, active suppression of fires, often with water, is all important. Fire suppression over a range of scales, from a single droplet to large sprinklers, is explored. The extinction process is shown using a unique line burner that slowly transitions to extinction as nitrogen is added to the surrounding flow. A dramatic intensification of combustion, fire whirls, is also presented, occurring when appropriate levels of swirl are added to a pool fire. They result in a dramatic increase in flame length, heat release, and, in nature, unpredictable fire behavior. The efficiency of this process and its potential application for remediation and energy production are presented through the "blue whirl," a soot-free regime of the fire whirl.

Larger uncontrolled fires, namely wildland fires, are also presented. In the laboratory, experiments using laser-cut fuelbeds have been studied in an effort to understand the processes controlling wildland fire spread. Also shown are firebrands, small burning embers that loft off burning material and cause much of the home destruction observed in wildfires. Large fires have also occurred in cities, and a recreation of the Great Fire of London of 1666 that occurred on the Thames in 2016 is a fantastic illustration of the destruction that has occurred in urban areas around the world.

Finally, the slow oxidation of carbonaceous material known as smoldering is presented. While it does not have a flame, the combustion process is critically important for biomass burning and the production of emissions from many different landscape-scale fires, impacting public health and the global climate.

As Michael Faraday once said, "There is no better, there is no more open door by which you can enter into the study of natural philosophy, than by considering the physical phenomena of a candle." This journey through fire presents only a brief introduction to many captivating features encountered in the study of fire phenomena.

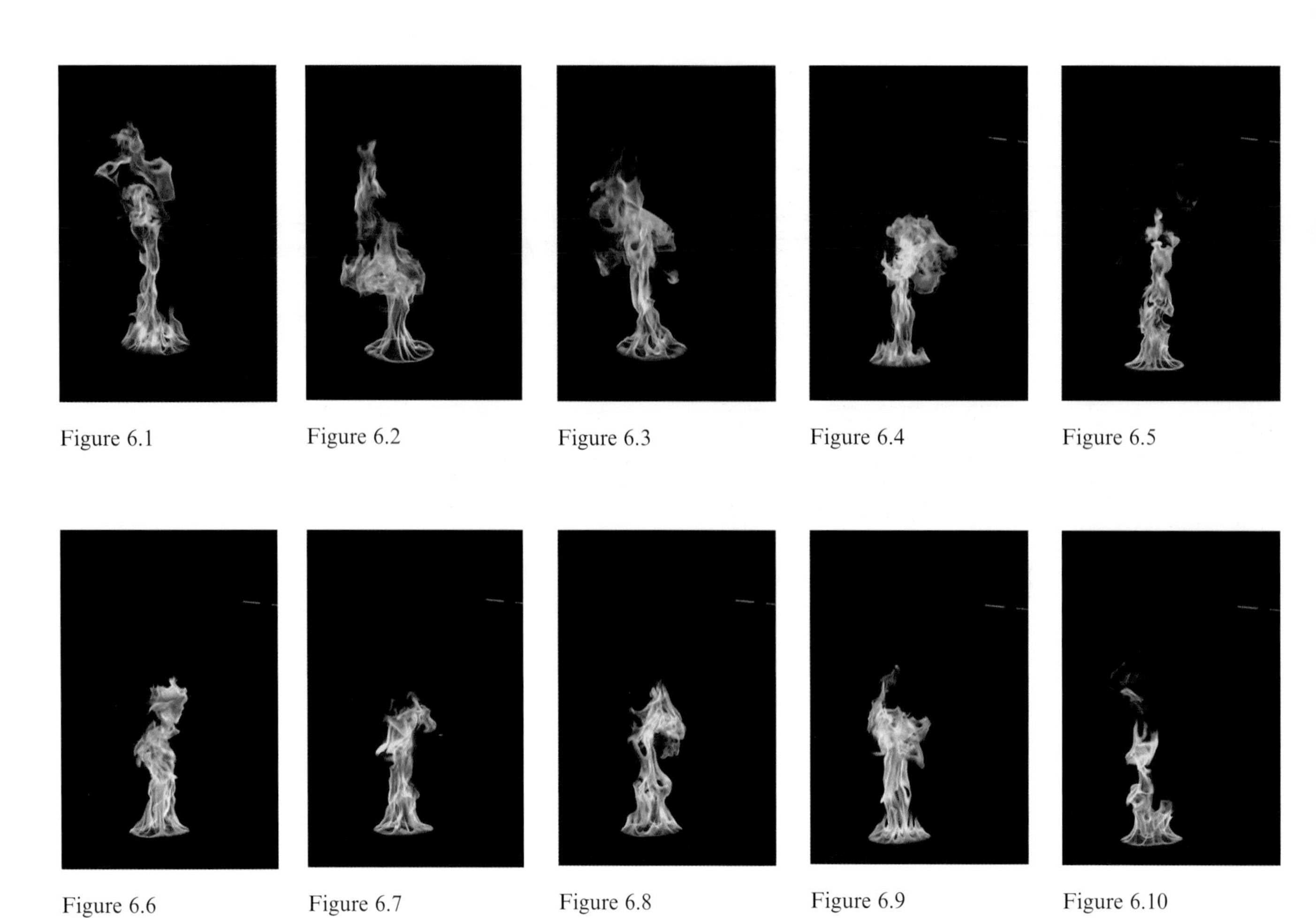

Figure 6.1 Figure 6.2 Figure 6.3 Figure 6.4 Figure 6.5

Figure 6.6 Figure 6.7 Figure 6.8 Figure 6.9 Figure 6.10

6.1 Pool Fires

6.1.1 Puffing Pool Fires

Carmen Gorska

University of Queensland

Buoyant diffusion flames, known as pool fires, form when a pool of liquid fuel is ignited. One interesting feature of pool fires is that they are known to "puff" at characteristic frequencies, on the order of 1 Hz, dependent on the diameter of the pool. In these photos two cycles of "puffing" are illustrated, where parcels of hot gas are formed at the base and transported upward, before the fire necks back in and the cycle begins again.

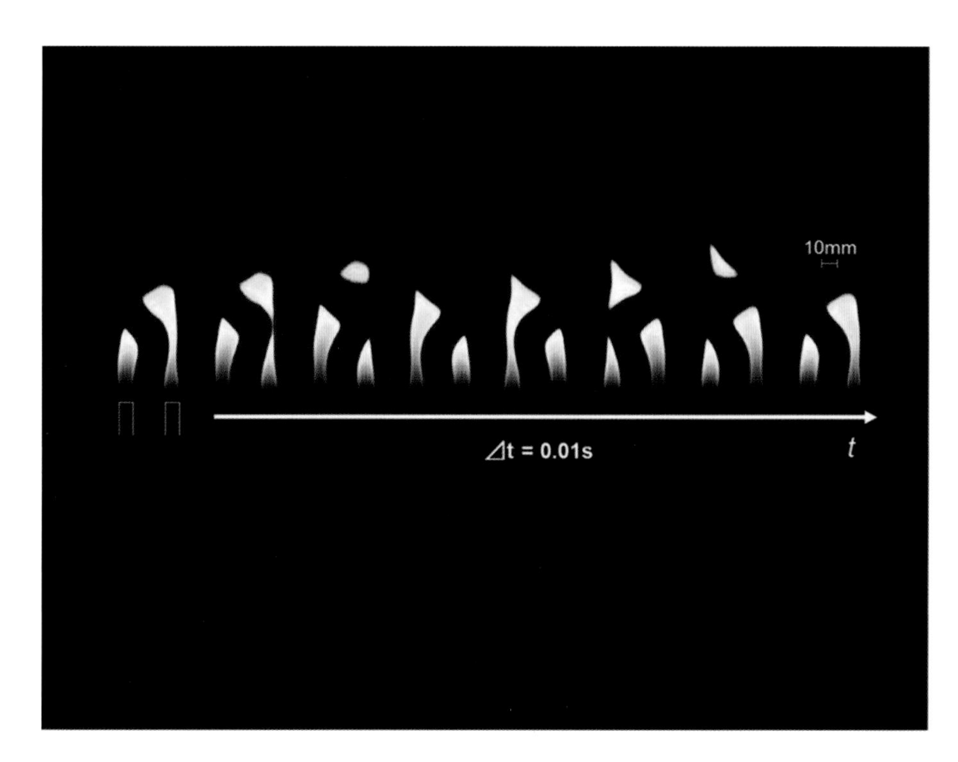

Figure 6.11

6.1.2 Synchronizing Twin Flickering Flames

Yuji Nakamura, Matsuoka Tsuneyoshi, and Keisuke Mochizuki
Toyohashi University of Technology

Diffusion flames at a certain scale under a gravimetric field produce periodic motions caused by a buoyancy-induced instability: this is what we call flickering. We examine what happens when twin flickering flames are placed at a prescribed distance from one another. Interestingly, both in-phase and out-phase harmonic oscillation of twin methane–air jet flames (10.6 mm inner diameter of burners, issuing methane at 500 cm^3/min) are found when the distance is modified. These images were taken with a high-speed camera (CASIO EX-F1 at 100 fps) at a distance of 25 mm, at which the out-phase mode with 13 Hz of flickering frequency is dominant. Left and right flames are elongated and their tips are separated in turns, making it look like the flames take turns jabbing at one another.

When the separation distance is set to smaller than 25 mm, twin flames merge together and show an in-phase oscillation. When it is larger than 25 mm, on the other hand, the oscillation mode suddenly switched to an out-of-phase one as shown in this figure. Interestingly, what is seen at 25 mm distance is that twin flames temporarily cease to suppress the buoyancy-induced instability together. In this way, we can study what causes this buoyancy-induced instability and how to suppress it if necessary.

Acknowledgements

This research was supported by Toyota Physical and Chemical Research Institute for fiscal year of 2015.

Keywords:
Flame flickering; buoyancy instability; synchronizing

Source Reference

1. K. Mochizuki, T. Matsuoka, Y. Nakamura, Study on oscillation and transition behavior of interacting flickering flames, *Bulletin of JAFSE (Japan Association for Fire Science and Engineering)* 67, 2 (2017) (in Japanese).

Other Relevant Reference(s)

2. K. Mochizuki, Y. Nakamura, Experimental study on dynamically synchronized behavior of two flickering jet flames, Proceedings of the 10th Asia-Oceania Symposium on Fire Science and Technology (AOSFST-10), Poster Session, Tsukuba, Japan (2015.10), FP-7.
3. Y. Nakamura, K. Mochizuki, T. Matsuoka, Interaction-induced flickering behavior of jet-diffusion flames, Proceedings of the 27th International Symposium on Transport Phenomena (ISTP-27), Hawaii, USA (2016.9), ISTP27–191.

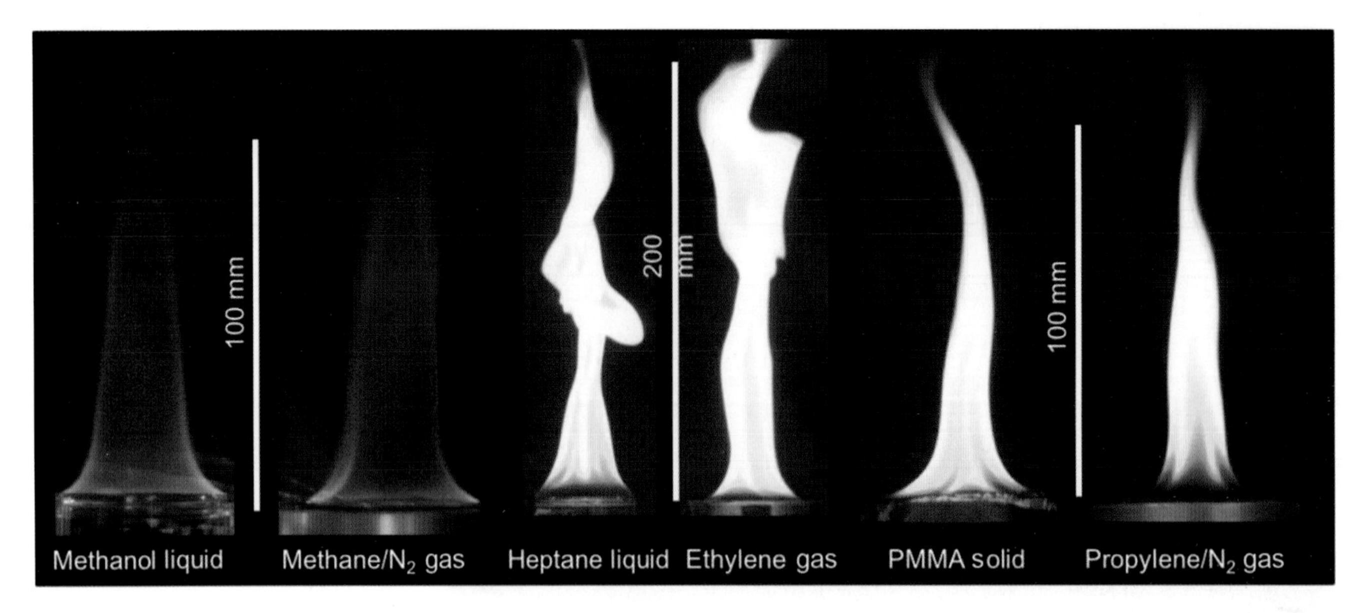

Figure 6.12

6.1.3 Emulation of Pool Fires with a Gas Burner

Yi Zhang, Peter Sunderland, and James Quintiere

University of Maryland

Studying the combustion of condensed fuels can be difficult, so researchers often attempt to "emulate" these flames. In each pair of images shown, the flames of condensed and gaseous fuels have the same heats of combustion, heats of gasification, surface temperatures, and laminar smoke points. When these four quantities are matched, the resulting diffusion flames have similar appearances. The burner used here has been designed to eventually study flames in microgravity.

References

Frida Vermina Lundström, Peter B. Sunderland, James G. Quintiere, Patrick van Hees, John L. de Ris, Study of ignition and extinction of small-scale fires in experiments with an emulating gas burner, *Fire Safety Journal*, Volume 87, 2017, Pages 18–24.

Yi Zhang, Matt Kim, Haiqing Guo, Peter B. Sunderland, James G. Quintiere, John deRis, Dennis P. Stocker, Emulation of condensed fuel flames with gases in microgravity, Combustion and Flame, Volume 162, Issue 10, 2015, Pages 3449–3455.

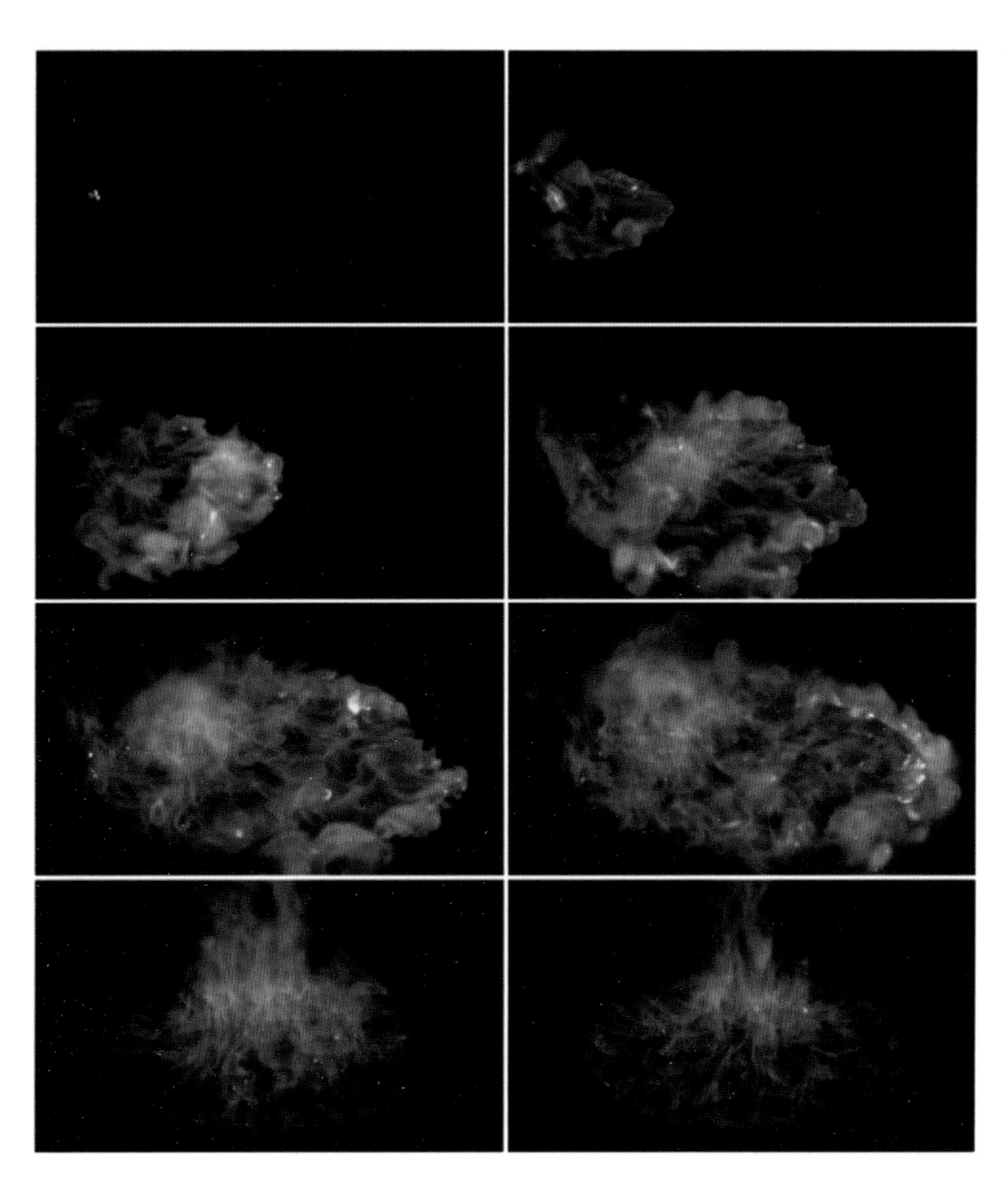

Figure 6.13

6.1.4 Methanol Pool Fire

Ian Grob

U.S. Forest Service

Pool fires have a complex structure, originating from the base upward into the plume. When burning methanol, little to no soot is produced, so the flame appears blue and structures can be clearly visualized. In this sequence of images, a pool of methanol ignited. The flame first is observed to propagate rapidly over the liquid fuel surface. Once the entire surface is ignited, a series of structures can be observed around the base of the pool fire, which play a complicated role in its structure and behavior.

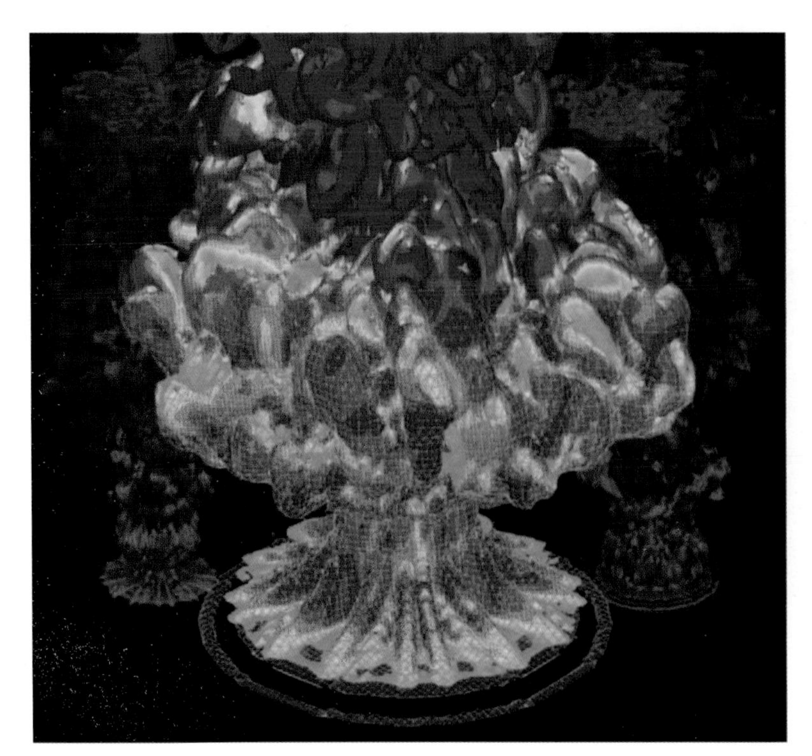

Figure 6.14

6.1.5 Simulation of Pool Fire Dynamics

Paul DesJardin

State University of New York at Buffalo

This graphic depicts the turbulent instability dynamics of a 1 m-diameter methane–air fire plume. Instability dynamics are responsible for the unsteady heat transfer in fire environments, which have been observed experimentally. The mesh superimposed on the bottom of the plume is the underlying computational grid utilized to carry out the calculation. The puff cycle of the plume may be broken down into four distinct stages: formation of a base instability near the edge of the plume; growth of the instability due to a misalignment of the vertical pressure and radial density gradients generating a localized torque; formation of a large toroidal vortex that self-propagates and entrains a large quantity of surrounding air; and destruction of the toroidal vortex due to formation of secondary instabilities that grow causing a nonlinear breakdown of the toroidal vortex. The baroclinic torque instability dynamics is an important ingredient in characterizing the fire dynamics of large pool fires as it dictates the level of air entrainment and flame surface area available for combustion. An improved understanding of instability dynamics will therefore result in more accurate predictions of fire intensity and growth.

Acknowledgements

This research was supported by the National Science Foundation under grant CTS-0348110 and the Office of Naval Research under Grant No. N00014-03-1-0369. Computer resources were provided by the Center from Computational Research (CCR) at the University at Buffalo, the State University of New York.

Keyword:

Turbulent fire plume, instability dynamics

Source References

1. P. E. DesJardin, Modeling of conditional dissipation rate for flamelet models with application to large eddy simulation of fire plumes, *Combustion Science and Technology* 177 (2005), 1881–1914.

Other Relevant Reference(s)

2. P. E. DesJardin, T. J. O'Hern, S. R. Tieszen, Large eddy simulations and experimental measurements of the near field of a large helium-air plume, *Physics of Fluids* 16 (2004), 1866–1883.
3. P. E. DesJardin, H. Shihn, M. D. Carrara, Combustion Subgrid Scale Modeling for Large Eddy Simulation of Fires, in *Transport Phenomena of Fires*, edited by B. Sunden and M. Faghri, WIT Press, Southampton, UK, 2008.

Figure 6.15 Largest experimental LNG pool fire on water (56 m-diameter, 146 m height) at Sandia National Laboratories, Albuquerque, NM.

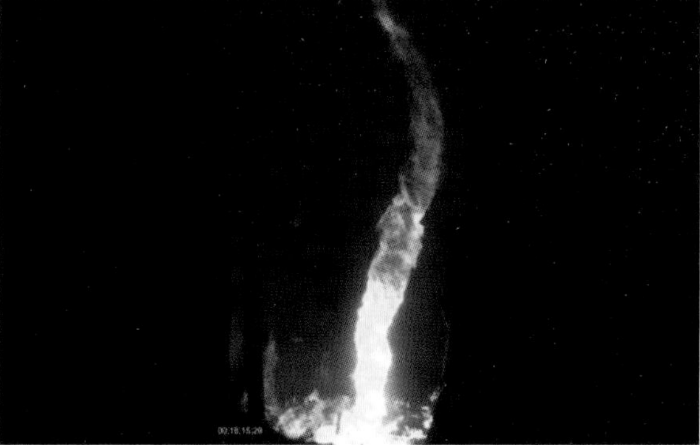

Figure 6.16 Largest indoor experimentally created fire whirl (3 m-diameter pool) at Sandia National Laboratories, Albuquerque, NM.

6.1.6 Extremely Large-Scale LNG Pool Fires and Indoor Fire Whirls

Anay Luketa

Sandia National Laboratories

While many experiments are conducted in the laboratory, it is important to verify and explore fire dynamics on a larger scale, where fire presents its greatest hazards. A large-scale liquefied natural gas fuel fire experiment simulating a spill of fuel on water was conducted at the 120 m-diameter water pool at Sandia National Laboratory, Albuquerque, New Mexico, on December 10, 2018. Other photos show an experimentally generated fire whirl formed from a 3-meter diameter JP-8 test conducted in the FLAME facility at Sandia National Laboratory.

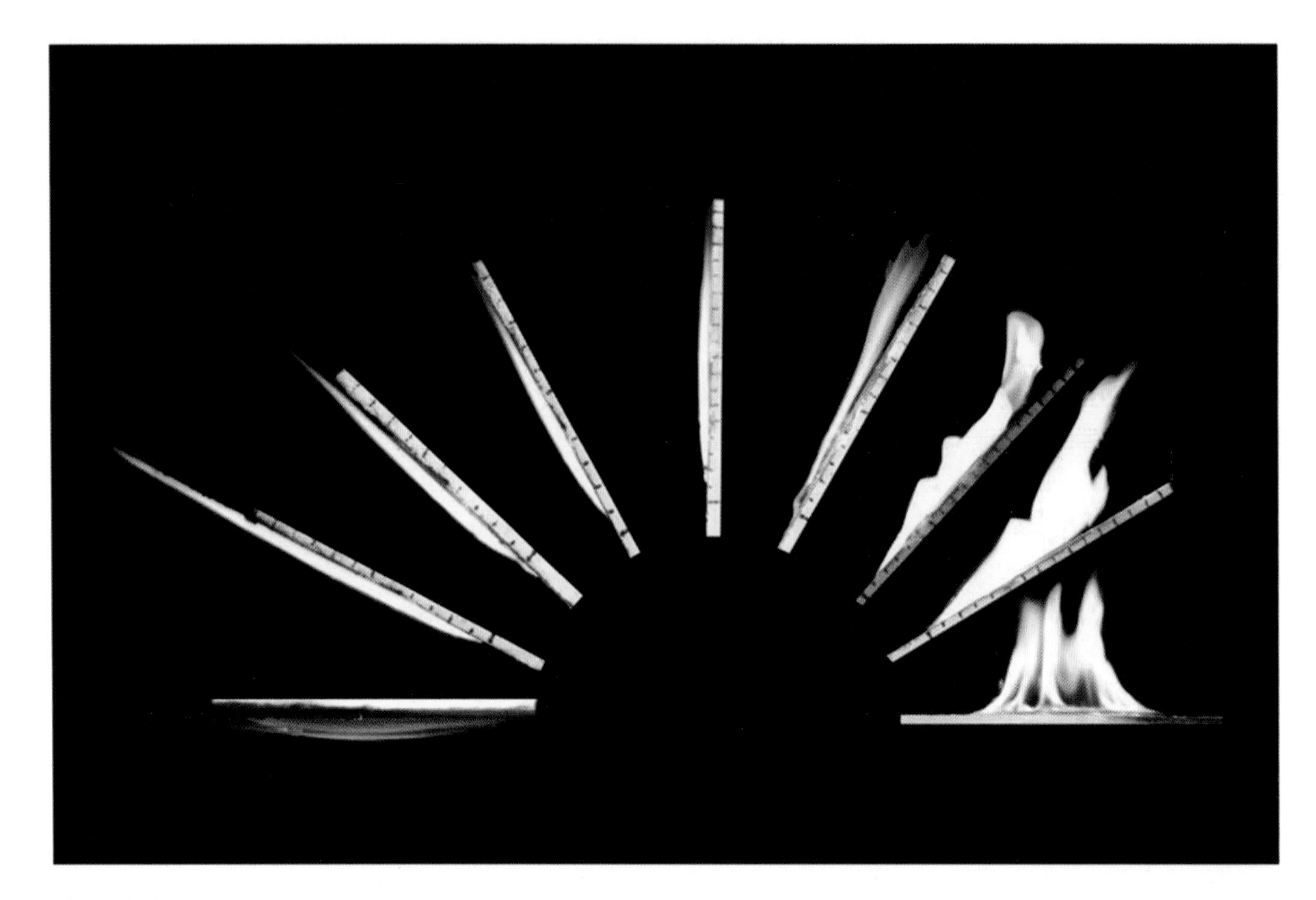

Figure 6.17

6.2 Flame Spread and Fire Growth

6.2.1 Surface Inclination Effects on Upward Flame Spread

Michael Gollner
University of California, Berkeley
Xinyan Huang
University of Maryland
Forman A. Williams and Ali S. Rangwala
University of California, San Diego

Flame spread, the process of fires moving over a surface, occurs due to ignition of virgin material as a result of heating. There are many mechanisms of flame spread, which vary depending on the materials, geometry, conditions, and scale under investigation. Changes in the rate of flame spread, and the nature of the flame, are easily visualized by changing the orientation of the fuel. Starting from the left "ceiling fire," as the inclination angle or tilt of a burning surface is increased, underside flames transition from blue, well-mixed laminar flames to increasingly turbulent yellow flames on the topside that "lift" from the surface, dramatically increasing the flame thickness. These images were taken perpendicular to the surface of a thick sample of Polymethyl Methacrylate mounted flush into insulation board as flames spread upward. These tests have helped in finding critical inclinations with maximum flame spread rates, burning rates, and heat fluxes from the flame.

Reference

Gollner, M. J., et al. "Experimental study of upward flame spread of an inclined fuel surface." *Proceedings of the Combustion Institute* 34.2 (2013): 2531–2538.

Figure 6.18

Figure 6.19

Figure 6.20

Figure 6.21

6.2.2 Wind-Driven Fires

Wei Tang, Colin Miller
University of Maryland

Michael J. Gollner
University of California, Berkeley

Changes in ambient wind, similar to slope, have a drastic affect on the spread rate of a fire. This series of images show the effect of wind (from 0.87 to 2.45 m/s) on a stationary flame, simulated using a 9.5 kW gas burner. As the wind, blowing from right to left, is increased, the fire transitions from more of a plume-dominated mode to an attached, boundary-layer flame. As flames become attached to the surface, heating and thus their rates of spread increase.

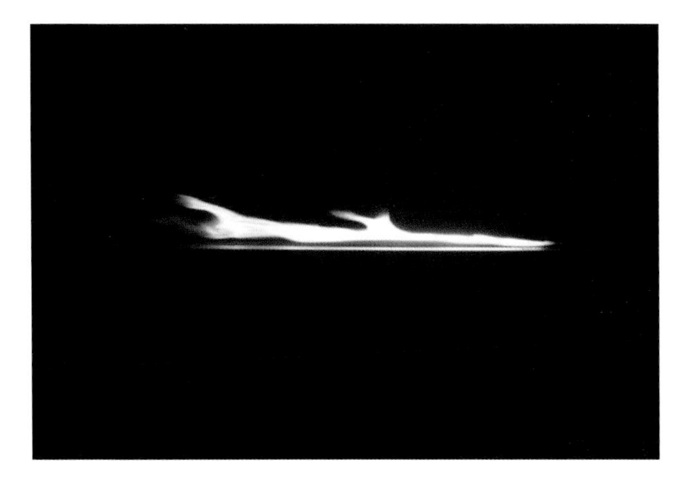

Figure 6.22

Reference

Wei Tang, Colin H. Miller, Michael J. Gollner, Local flame attachment and heat fluxes in wind-driven line fires, Proceedings of the Combustion Institute, Volume 36, Issue 2, 2017, Pages 3253–3261, ISSN 1540-7489, https://doi.org/10.1016/j.proci.2016.06.064.

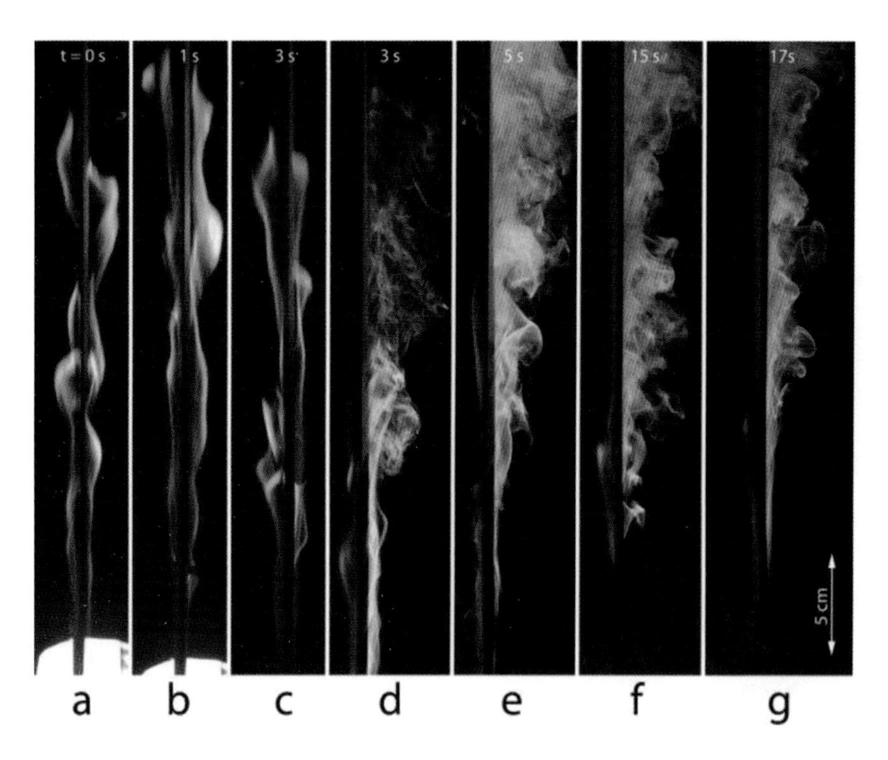

Figure 6.23

6.2.3 Asymmetric Buoyancy Induced Blow-Off

Michael C. Johnston, James S. T'ien, Derek E. Muff, and Xiaoyang Zhao

Case Western Reserve University

Sandra L. Olson and Paul V. Ferkul

NASA Glenn Research Center

A 5 cm-wide × 20 cm-long composite fabric sheet woven with thread made of 75% cotton and 25% fiberglass is hung vertically and is electrically ignited at the bottom surface to cause a flame to spread upward. When burned, the cotton is pyrolyzed into fuel vapor leaving behind a woven fabric made only of fiberglass. This remaining fiberglass maintains the integrity of the fabric and acts as a flame barrier that does not allow ignition to penetrate it.

The image views the hanging fabric from the narrow edge. As the flame grows in size after ignition, the incoming air velocity induced by buoyancy becomes too high to remain stable. Image (a) shows the maximum flame length of about 20 cm after ignition and initial flame growth. (b) The flame base on the right-hand side is beginning to blow off due to the high velocity of the entrained air. (c) The flame base on the right-hand side continues to blow off, while the flame base on the left-hand side is maintained. Because the fuel is thermally thin, the left-hand side flame continues to vaporize fuel on the right-hand side, seen as white smoke. (d) The right-hand side flame has been blown too far downstream. The material surface is still too cold in this region for the flame base to be stable in this position. (e) The right-hand flame has completely blown off. A large amount of fuel continues to vaporize from the surface. (f) Due to the reduced fire power and heat feedback to the fuel, the left-hand flame shrinks in length. (g) The left-hand flame reaches its final stable length. As the cotton fuel burns out, this flame slowly propagates upward across the entire surface toward new fuel, but at this reduced length of about 10 cm. The remaining flame will not blow out due to the reduced entrained velocity. The fuel vapor mixing itself with air on the right-hand side is still flammable but is not reignited by the left-hand flame because of the flame barrier properties of the remaining fiberglass fabric.

Acknowledgements

This research was supported by NASA and Underwriters Laboratories.

Keywords:

Buoyant blow-off; upward burning limit; flammability; flame spread

Reference

1. M. C. Johnston, S. James, D. E. Muff, X. Zhao, S. L. Olson, P. V. Ferkul. Self induced buoyant blow off in upward flame spread on thin solid fuels. *Fire Safety Journal* 71 (2015), 279–286.

Figure 6.24

6.2.4 Small-Scale Corner Fire

Nils Johansson
Lund University

The photo shows a turbulent diffusion flame from a $0.1 \times 0.1\ m^2$ sandbox propane burner placed in a corner in a 1/3-scale ISO 9705 room. The heat release rate was measured with the oxygen depletion method to approximately 30 kW. The test setup is used in educational and research activities at the Division of Fire Safety Engineering at Lund University, Sweden.

Corner fires are used to evaluate the fire characteristics of combustible lining materials in standardized tests like ISO9705. Fires that interact with a wall or corner with combustible linings will in most cases result in a more rapid fire spread and shorter time to flashover than if no such interactions are present. The heat flux to the linings will be due to both the exposure fire and the room environment. Furthermore, the mean flame height will be longer when the fire is next to a wall or in a corner because less air can be entrained; this results in the fuel having to travel a longer distance to become fully combusted.

Acknowledgements

Stefan Svensson, Reader at the Division of Fire Safety Engineering, is gratefully acknowledged for his assistance when the test was performed.

Keywords:

Corner fire; propane; diffusion flame; enclosure effects

Figure 6.25

Figure 6.26

Figure 6.27

Figure 6.28

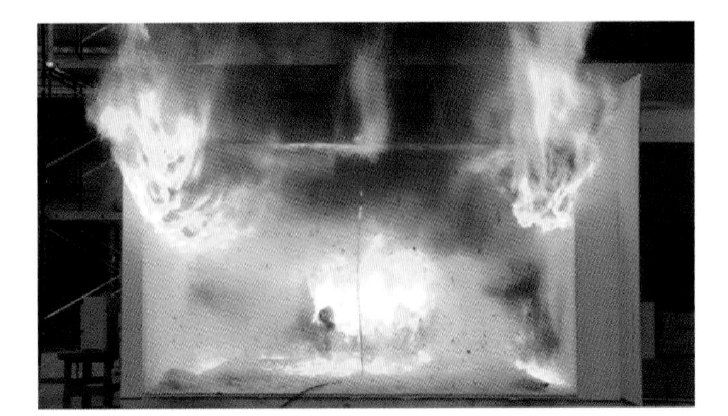

Figure 6.29

Figure 6.30

6.2.5 Progression of a Compartment Fire

Mark Wahl and Jonathan Butta

ATF Fire Research Laboratory

An oversized chair, consisting of mostly polyurethane, was ignited by an FRL "Standard Ignition Package." The chair was placed into a flashover cell that contained only carpet at the floor with gypsum wallboard walls

Figure 6.31

Figure 6.32

Figure 6.33

and ceiling. The fire on the chair was allowed to grow, eventually developing a thermal layer at the ceiling sufficient for flashover conditions. Figures 6.26–6.34 are a series of images demonstrating the development of this compartment fire.

Keywords:
Compartment; fire; flashover; rollover; ignition; furniture

Figure 6.34

6.2.6 Melting and Dripping during Flame Spread over Wires

Xinyan Huang, Andy Rodriguez, and Carlos Fernandez-Pello

University of California, Berkeley

Dripping of molten fuel can change fire behaviors, ignite nearby materials, and expand fire size. Dripping occurs when gravity overcomes the resistance from surface tension and viscous forces. It involves a complex phase-change process and is most common in fires of electrical wires and heat insulation materials. The image shows the dripping of molten polyethylene insulation in an electrical wire fire. Dripping first ignited the sand soaked with alcohol, and then started to frolic (interact) with the puffing pool fire. The whole process lasted less than 0.2s, and was captured by a high-speed camera (500 fps).

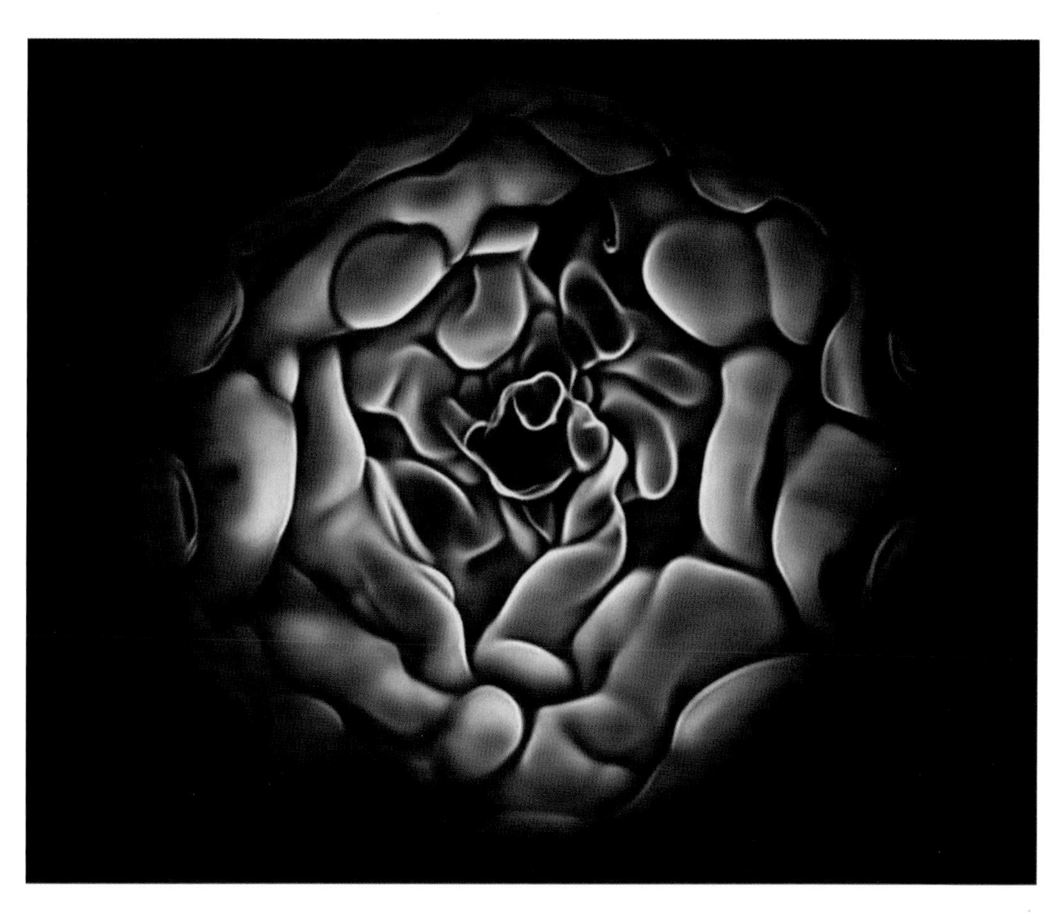

Figure 6.35

6.2.7 Ceiling "Blooming" Fire

Longhua Hu
University of Science and Technology of China

The image presented is an experimental flame photo captured from a propane fire beneath the ceiling generated from a stainless circular burner (diameter: 0.06 m, gas exit velocity: 0.015 m/s, heat release rate: 15.3 kW, source Reynolds number: 203). The circular burner was installed at the center of the ceiling with the outlet downward and flush with the ceiling. The ceiling was made from mica board with a low thermal conductivity of 0.035 W/(m K) and good thermal resistance performance. The fuel gas was discharged downward with pure propane as the fuel and air as the oxidizer, which is diffusing with fuel to produce a diffusive flame. The entire apparatus was located in a still air environment at standard pressure. The picture is obtained by a CCD camera with 1/180 s exposure time. This beautiful "blooming flower" flame is broken up into "petals" whose sizes increase during "blooming" from the center. Such gorgeous "blooming fire" is not only esthetically attractive but also a physical nature of buoyant thermal-diffusive instability, whose characteristic length scale can be well viewed from the "petal" sizes. It is this kind of beauty that drives our fascination with the art of fire.

Acknowledgements

This work was supported jointly by Key project of National Natural Science Foundation of China (NSFC) under Grant No. 51636008, Excellent Young Scientist Fund of NSFC under Grant No. 51422606, the Newton Advanced Fellowship (NSFC: 51561130158; RS: NA140102), Key Research Program of Frontier Sciences, Chinese Academy of Science (CAS) under Grant No. QYZDB-SSW-JSC029, Fok Ying Tong Education Foundation under Grant No. 151056, and Fundamental Research Funds for the Central Universities under Grant No. WK2320000035.

Keywords:

Ceiling; blooming fire; propane; circular burner; thermal; diffusive instability

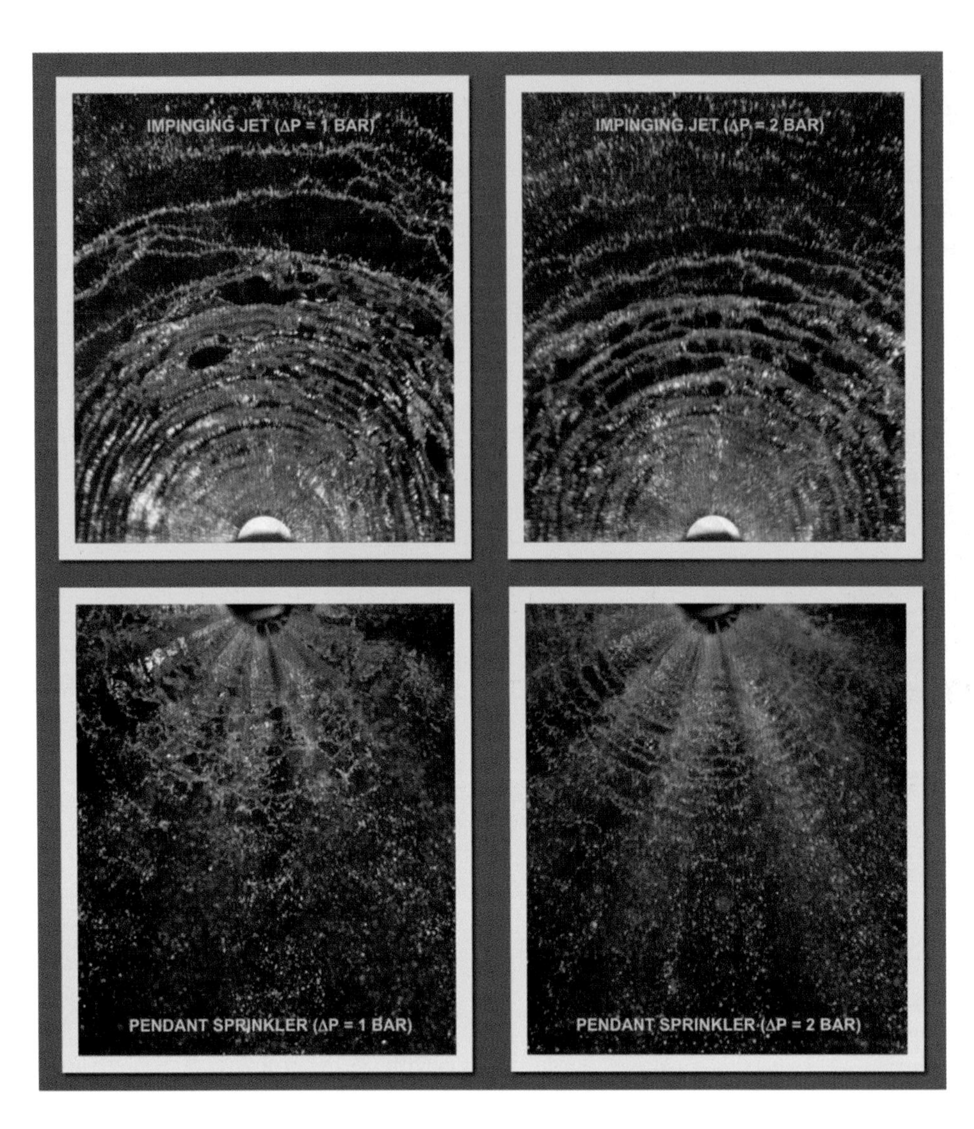

Figure 6.36

6.3 Fire Suppression

6.3.1 Sprinkler Discharge Characteristics

Ning Ren, Andrew F. Blum, Ying-Hui Zheng, Chi Do and André W. Marshall

University of Maryland

In order to extinguish a fire, oxygen, heat, or fuel must be removed from the combustion process. A variety of techniques have been developed to extinguish fires, most notably water-based fire sprinklers. Sprinklers, however, use complex fluid dynamics to atomize a stream of liquid water into a spray that can reach a fire plume and, often, the solid fuel underneath. Here, essential sprinkler atomization features are revealed in the canonical impinging jet configuration ($K = 26\ \text{lpm/bar}^{0.5}$). A thin unstable sheet is formed along the deflector. The sheet breaks up into ligaments that disintegrate into drops. The same atomization mechanisms are observed in a pendant sprinkler configuration with similar geometry; however, the tines introduce 3D effects.

Reference

Ren, N., Blum, A., Zheng, Y., Do, C. and Marshall, A., 2008. Quantifying the Initial Spray from Fire Sprinklers. *Fire Safety Science* 9: 503–514. doi:10.3801/IAFSS.FSS.9-503.

Figure 6.37

6.3.2 Oxidizer Dilution Quenching of a Turbulent, Methane Line Flame

James P. White, Eric D. Link, Taylor M. Myers, Andre W. Marshall, and Peter B. Sunderland
University of Maryland

Nitrogen gas (99.99% pure) is added to a co-flowing oxidizer stream of ambient air at atmospheric temperature and pressure to accomplish variations in the oxidizer oxygen concentration. The co-flowing oxidizer surrounds a 50 cm-wide by 50 cm-long stainless-steel slot burner flowing methane gas (99.5% pure) at 1.0 g/s to yield a nominal 50 kW turbulent diffusion flame with a mean visible flame height of roughly 50 cm. Image slices from left to right depict flame response to decreasing oxidizer oxygen concentration. From left to right, image slices correspond to oxidizer oxygen concentrations of 20.9 vol% O_2, 17.5 vol% O_2, 15.6 vol% O_2, 14.8 vol% O_2, 13.8 vol% O_2, and 13.2 vol% O_2.

With reducing oxidizer oxygen concentration, soot production is inhibited and flame temperature is reduced, leading to a transition in flame color from bright yellow to pale blue. Here, yellow flame regions are made visible by the incandescence of soot particles, and blue flame regions are made visible by the relatively dim luminescence of CH radicals. The dim blue regions are visible only in the absence of the comparably bright soot incandescence.

Acknowledgements

Support for this work has been provided by the U.S. National Science Foundation (NSF-GOALI Award #1236788). Additional support provided by FM Global and United Technologies Research Center is gratefully acknowledged.

Keywords:
Flame suppression; extinction; oxidizer dilution; soot; incandescence

Source References

1. J. P. White, E. D. Link, T. M. Myers, A. W. Marshall, P. B. Sunderland, Oxidizer dilution quenching of a turbulent, methane line flame, Fire Safety Science – Proceedings of the Eleventh International Symposium, Canterbury, New Zealand, February 2014, image poster.

Other References

2. J. P. White, E. D. Link, A. Trouvé, P. B. Sunderland, A. W. Marshall, A general calorimetry framework for measurement of combustion efficiency in a suppressed turbulent line fire, *Fire Safety Journal* 92 (2017), 164–176.
3. J. P. White, E. D. Link, A. Trouvé, P. B. Sunderland, A. W. Marshall, J. A. Sheffel, et al., Radiative emissions measurements from a buoyant, turbulent line flame under oxidizer-dilution quenching conditions, *Fire Safety Journal* 76 (2015), 74–84.
4. J. P. White, S. Verma, E. Keller, A. Hao, A. Trouvé, A. W. Marshall, Water mist suppression of a turbulent line fire, *Fire Safety Journal* 91 (2017), 705–713.

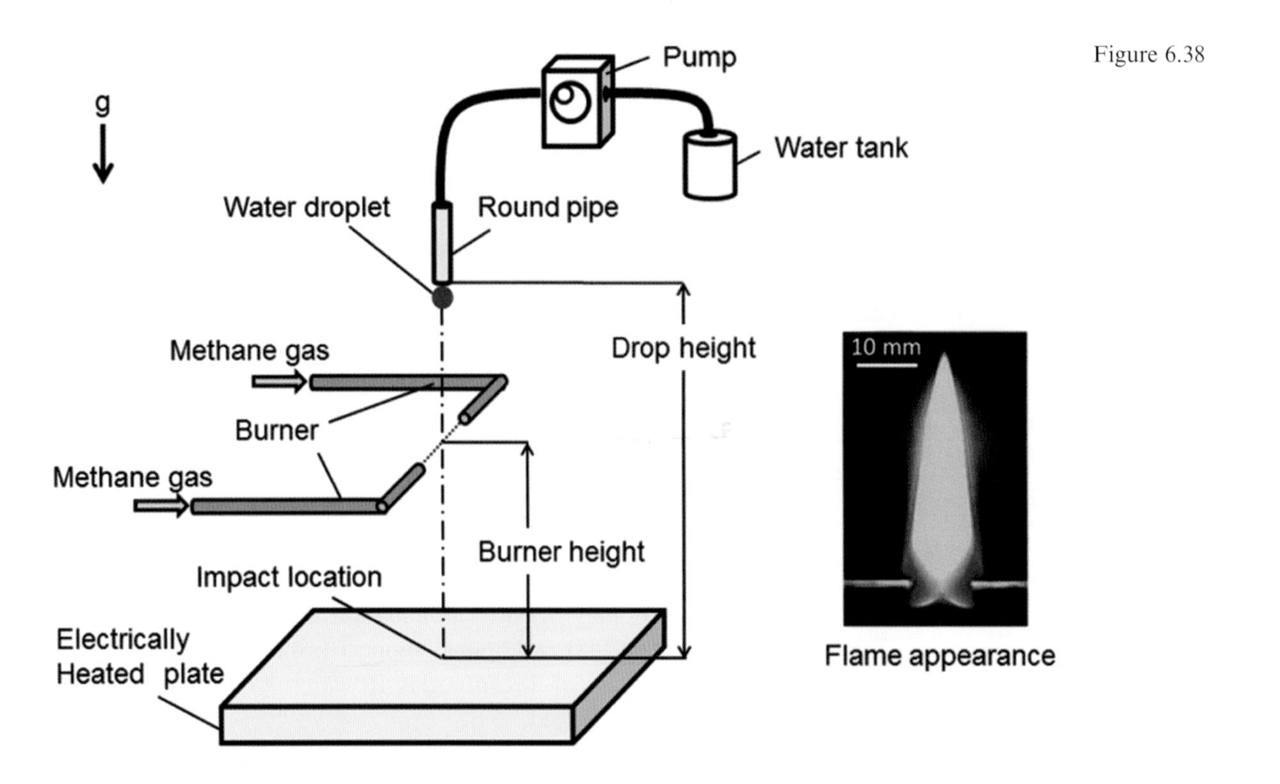

Figure 6.38

6.3.3 Indirect Extinguishment of a Diffusion Flame with Water Vapor

Hiroyuki Torikai

Hirosaki University

Figure 6.39 shows the experimental apparatus to extinguish a methane–air diffusion flame with water vapor produced by impacting a single water droplet onto an electric heated plate. The plate surface temperature was varied from 100°C to 430°C. The water droplet whose diameter was about 3.2 mm was dropped from a height of 400 mm. The diffusion flame was formed with a horizontally-opposed tube burner, whose outer and inner diameters were 3 mm and 2.6 mm, respectively. The distance between the burner exits was 10 mm. The volumetric flow rate of methane gas was set at 0.23 l/min. The flame appearance is shown in Figure 6.39, and the flame height was about 40 mm. The distance from the burner to the heated plate surface (burner height) was varied from 32 to 102 mm. Figure 6.40 shows a sequential images of the impacting and boiling behaviors of the water droplet on the plate heated at 250°C, which were recorded by using a high-speed camera (exposure time: 1/10,000 s; frame rate: 10,000 fps) and a metal halide lamp. At 4.0 ms after the droplet impact, the shape of the water droplet deforms into the axisymmetric circular thin liquid film. At 5 ms, nucleate boiling begins in the thicker liquid layer at the outer edge of the circular water film. Then, at between 5.5 ms and 7.5 ms, the ring shape of the nuclear boiling region spreads inwardly to the center point of the liquid film. At 8 ms, spreading of the boiling region reaches the center of the liquid film. After that, the liquid film breaks up into droplet particles with various diameters. Figure 6.41 shows the extinguishing process of the flame at the plate temperature of 250°C and the burner height of 72 mm. The water vapor is basically invisible, but the water fog formed by condensation of the water vapor due to cooling is visible. Therefore, the motion of the visible water fog is considered to be almost the same as the water vapor behavior. The images were recorded with a digital camera (exposure time: 1/2,000 sec; frame rate: 300 fps). It is found that at 57 ms in Figure 6.41, the water vapor vortex ring is formed from the impacted water droplet. Then the vortex ring travels upward to the flame. At 113 ms, the vortex ring reaches the flame base, and the local extinction at the flame base region occurs at 127 ms. After that, the luminous flame area is blown downstream and finally the extinguishment is achieved. Thus, the produced water vapor is enough to extinguish the flame and

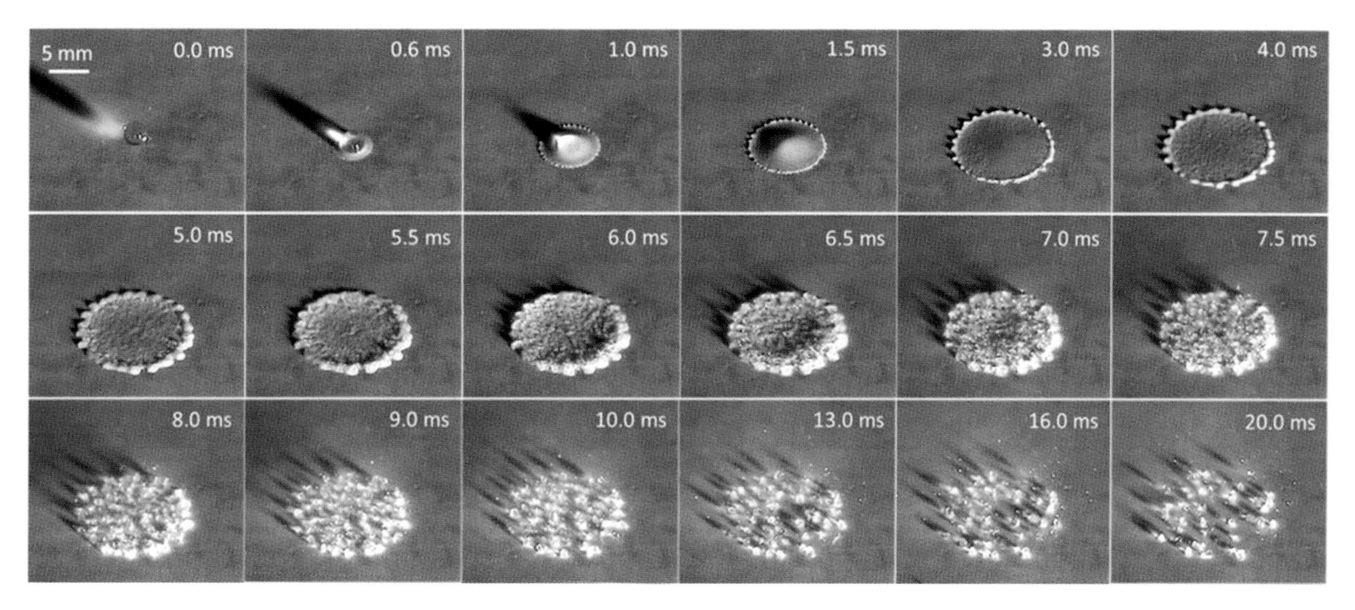

Figure 6.39

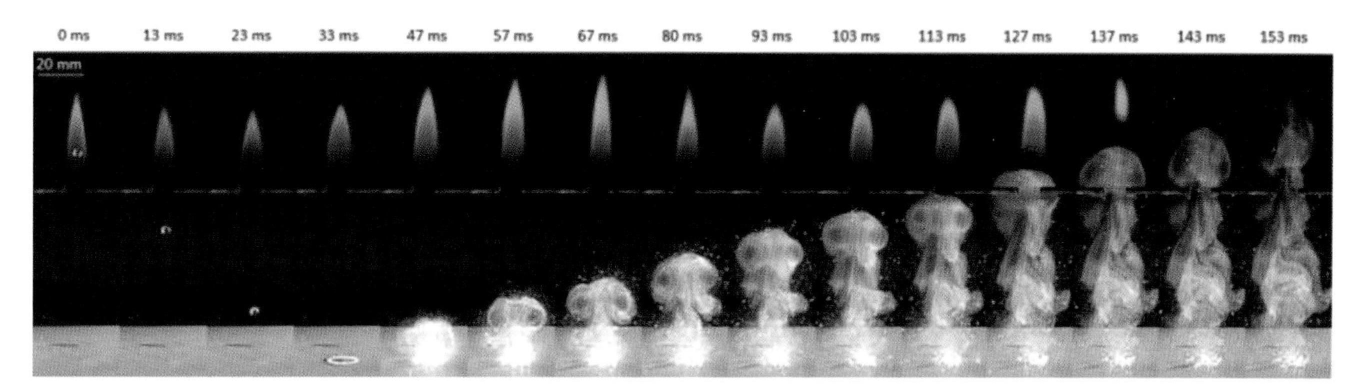

Figure 6.40

to be transported effectively to the combustion zone by the vortex ring.

In the confined fire space, there are high-temperature solid surfaces such as wall, ceiling, and floor heated by flames and fire plumes and pyrolysis region of the flammable materials. When the large water droplet is impinged onto the high-temperature solid surface in the burning area, it can be expected that the water droplet absorbs heat from the solid surface and turns rapidly into water vapor, and by producing a large amount of water vapor, fire extinguishment may be achieved. Based on the above considerations, we have investigated an extinguishing method with water vapor caused by impacting large water droplets onto the headed solid surface, which is called an indirect fire attack method.

Source Reference

H. Torikai and A. Ito, "Extinguishing characteristics of a diffusion flame with water vapor produced from a water droplet impacting onto a heated plate," *Fire Safety Journal*, in press (2017). https://doi.org/10.1016/j.firesaf.2017.03.012

Other Reference

Y. Chiba, H. Torikai and A. Ito, "Extinguishment Characteristics of a Jet Diffusion Flame with Inert-Gas Vortex Ring," *Progress in Scale Modeling* Vol. 2, pp. 115–125, 2014.

M. Ishidoya, H. Torikai and A. Ito, "Examination of Extinguishment Method with Liquid Nitrogen Packed in a Spherical Ice Capsule," *Fire Technology* Vol. 52, Issue 4, pp. 1179–1192, 2016.

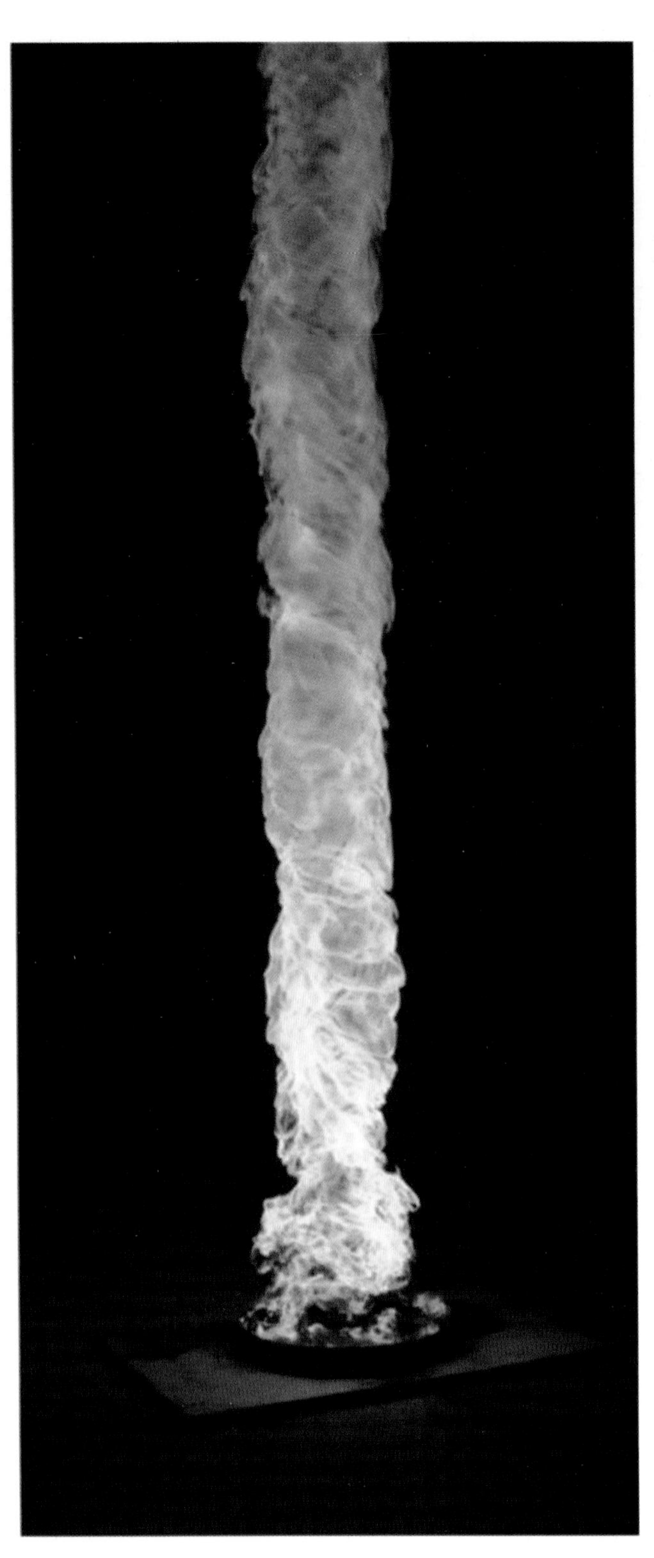

Figure 6.41

6.4 Fire Whirls

6.4.1 Fire Whirl

Sriram Bharath Hariharan, Ali Tohidi
University of Maryland

Michael J. Gollner
University of California, Berkeley

Fire whirls are one of the most destructive yet fascinating fire phenomena. They are often observed in large-scale urban and wildland fires and have caused incredible destruction due to their intensification of combustion and ability to spread fires during large conflagrations. This figure shows the full structure of a 30 cm-diameter heptane pool fire whirl created in the laboratory at the University of Maryland. The fire whirl is generated by adding swirl through four tangential slits around the edge of the walled test region.

Reference

Tohidi, Ali, Michael J. Gollner, and Huahua Xiao. "Fire Whirls." *Annual Review of Fluid Mechanics* 50, 187–213, 2018.

Figure 6.42

6.4.2 Fire Whirls from Multiple Sources

Egle Rackauskaite, Xinyan Huang, and Guillermo Rein

Imperial College London

Fire whirls often occur in large-scale urban and wildland fires. They are disastrous and capable of destroying everything in their paths. This picture shows laboratory-scale 30 cm-tall fire whirls generated by three identical pool fires placed at three different heights. These pool fires are placed inside a rectangular glass enclosure (35 × 35 × 60 cm) with an open top and 5 cm gaps placed at each of the four corners. Each pool first produces an individual long but unstable rotating fire. After a transition of about 30 s, the three pool fires start braiding into a single strong fire whirl as shown in the image, and consume the liquid fuel 10 times faster than the original individual pool fires.

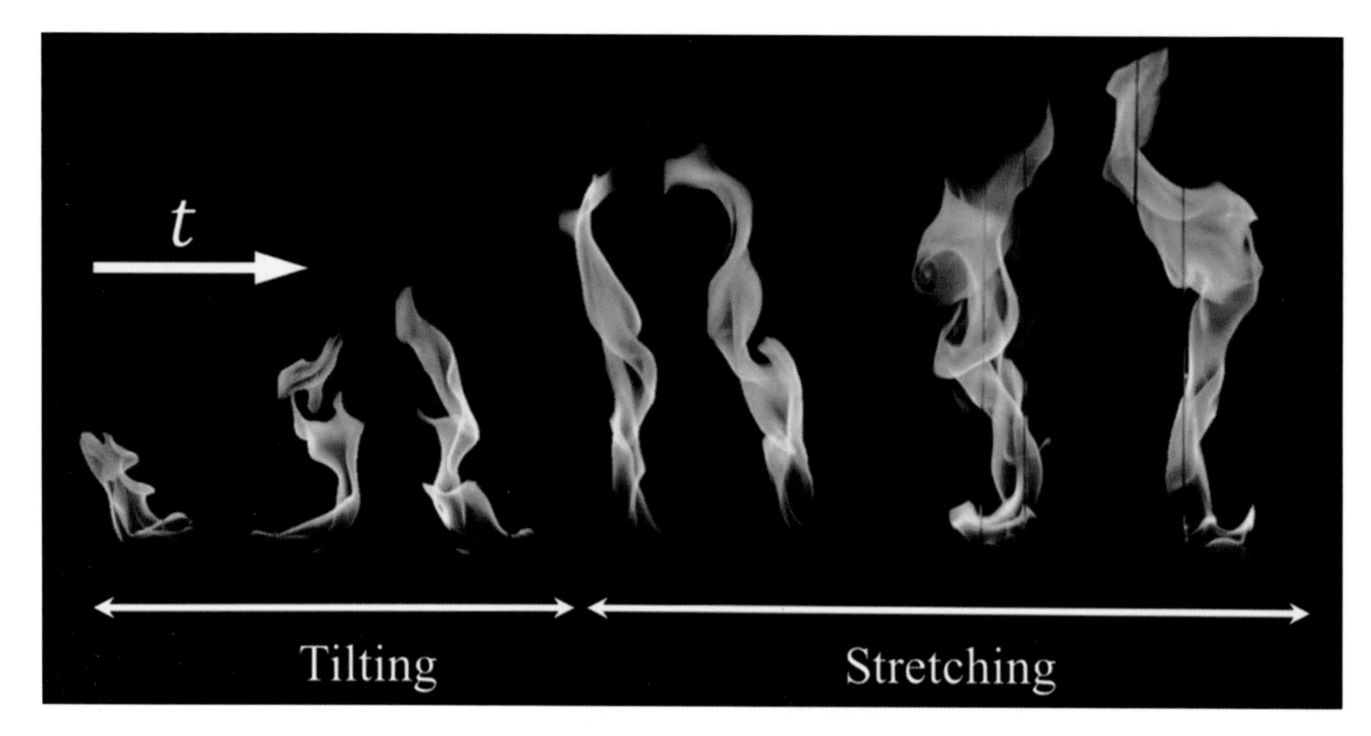

Figure 6.43

6.4.3 Formation of a Fire Whirl

Sriram Bharath Hariharan, Ali Tohidi
University of Maryland

Michael Gollner
University of California, Berkeley

Fire whirls require specific conditions and wind and fire to form. These conditions are often met in the laboratory via sidewalls with specifically placed slits, but in nature, topographical barriers, wind, and even the fire itself can generate the required circulation to form a central vortex core that results in this fascinating intensification of combustion. During the formation process, a traditional pool fire transitions once acted upon by external circulation, resulting in tilting and stretching, as shown below, until a stable vortex core is formed.

Reference

Tohidi, Ali, Michael J. Gollner, and Huahua Xiao. "Fire Whirls." *Annual Review of Fluid Mechanics* 50, 187–213, 2018.

Figure 6.44

Figure 6.45

6.4.4 The Blue Whirl

Sriram Bharath Hariharan, and Elaine Oran

University of Maryland

Michael Gollner

University of California, Berkeley

The blue whirl is a regime of the fire whirl that produces no soot, even while burning liquid hydrocarbons directly. In this experiment, a fixed-frame self-entraining fire whirl setup over a water surface was used to create a traditional yellow fire whirl using heptane. The fire whirl (~70 cm in height) was then observed to develop into a small blue whirl ~5 cm in height.

In this image, the recirculation within the blue cone region can be visualized using the soot particles, some of which are seen leaving the whirl through the top.

The recirculating regions are typical of the bubble mode of vortex breakdown, although a complete understanding of the flame regime does not exist yet.

This image was captured using a digital camera at f/2.8, 1/25 s, ISO-800.

Acknowledgements

This research was supported by NSF CBET-1507623, CBET-1554026, Minta Martin Endowment Funds in the Department of Aerospace Engineering, and Glenn L. Martin Institute Chaired Professorship at the A. James Clark School of Engineering.

Keywords:

Blue whirl; fire whirl; vortex breakdown

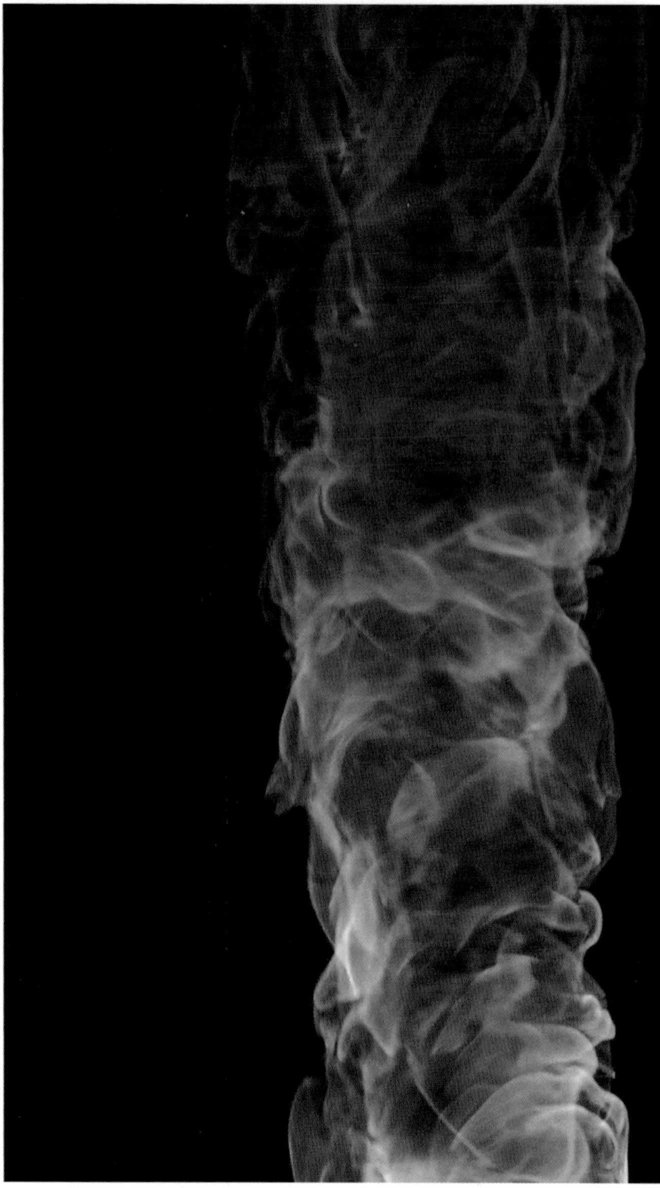

Figure 6.46

6.4.5 Inside the Whirl

Ian Grob

U.S. Forest Service

Fire whirls involve complex flame structures that have not yet been fully understood. These images were taken at the US Forest Service, Missoula Fire Sciences Laboratory where fire whirls are studied as they relate to hazards in large wildland fires. A close shot of the many flame structures in a vortex core are visible on the left, while a shot from above the fire whirl looking down the "donut hole" created by the whirl reveals the swirling motion in the center of the vortex core.

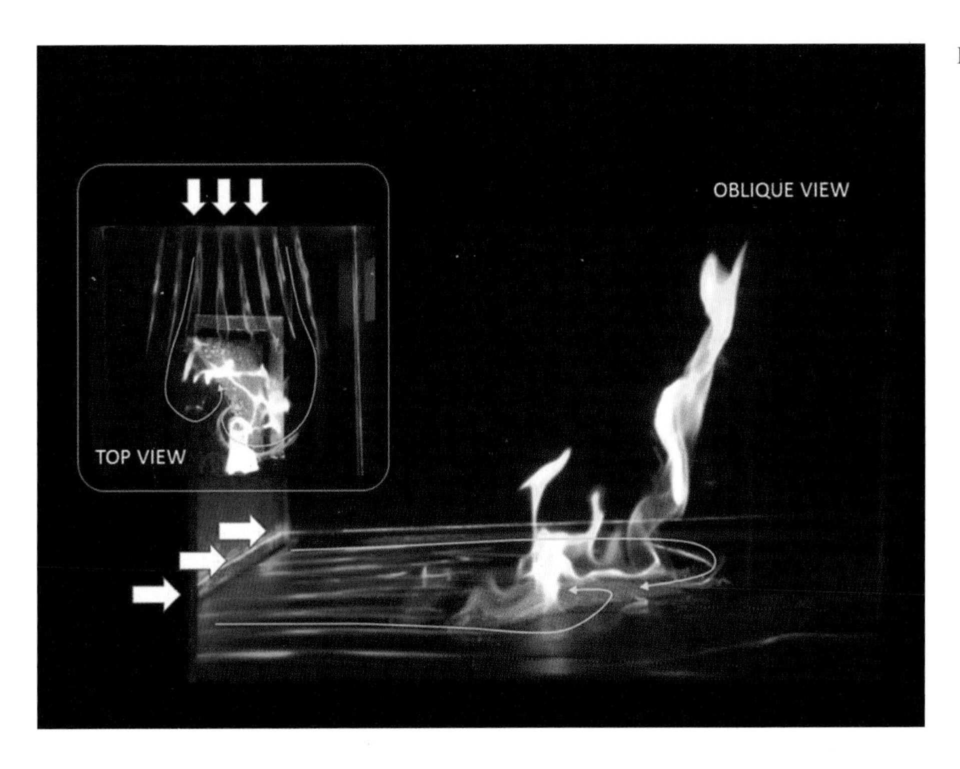

Figure 6.47

6.4.6 Twin Fire Whirls over a Methanol Liquid Pool in a Small Wind Tunnel

Yuji Nakamura, Matsuoka Tsuneyoshi, Shiino Kota, and Mizuno Mikio

Toyohashi University of Technology

This image shows twin fire whirls formed over a liquid methanol pool mounted in a small wind tunnel (both top and oblique views). An L-shaped tray 20 mm wide filled with ethanol was embedded in the floor of the wind tunnel, and a triangle shape of cotton sheet soaked in ethanol was placed inside the L-shape. A specially designed wind tunnel was used, which has been developed for a scale model experiment of the fire and installed at Toyohashi U Tech. The fuel tray was water-cooled, and its fuel level could be maintained during the test event. A uniform flow of 0.4 m/s was applied to the fire field, and soon twin fire whirls appeared as shown. The main fire whirl that appeared at the edge of the fuel tray showed the clockwise spin, while the small fire whirl that appeared on the inside of the L-shape showed the counter-clockwise spin. Twin vortexed flames can be nicely arranged to show counter-vortex formation in the fire field.

This experiment aimed to examine the location in which fire whirls may appear under various environmental conditions (e.g., flow velocity, tray size). This test facility provided steady conditions so that the location of twin fire whirls under the prescribed condition could be well explored. Scale model test results were compared to a 5 m scaled fire whirl test performed at Building Research Institute (BRI), Japan, and it was found that the potential location of the fire whirl could be predicted well by our scale model experiment.

Acknowledgements

This research was partially funded by Takahashi Industrial and Economic Research Foundation (FY2013), JKA (FY2014-2015), The Hibi Foundation (FY2015), and Grant-in-Aid for Challenging Exploratory Research (#15K12472) (FY2015-2016). Authors appreciate fruitful discussions with Mr. Sekimoto from Sekimoto PE., Japan, and Prof. Kazu Kuwana from Yamagata University, Japan.

Keywords:
Fire whirl; scale model experiment; wind tunnel; counter; vortexes

References

1. Y. Nakamura, K. Shiino, T. Nakashima, How well we can predict the occurrence of large fire whirl through scale model experiment?, Proceedings of the 12th International Conference on Flow Dynamics (ICFD2015), Sendai Japan (2015.11), (invited) OS3–3.
2. K. Shiino, Y. Nakamura, Mobility of fire whirl formed over l-shape ethanol tray embedded in small-scale wind tunnel, Proceedings of the 10th Asia-Oceania Symposium on Fire Science and Technology (AOSFST-10), Poster Session, Tsukuba, Japan (2015.10), FP-8.

Figure 6.48

6.5 Wildland Fires

6.5.1 Wildfires

Ian Grob

U.S. Forest Service

Wildfires burn somewhere around the globe every day, destroying vegetation and property and producing significant emissions into the atmosphere. While they can be incredibly destructive, fires are also an essential process for many ecological systems, occurring for millennia. Here, the most intense type of wildland fire, a crown fire, is shown as part of the crown fire experiments performed in the Northwest Territories, Alberta, Canada. This prescribed fire experiment included several experimental fire shelters used to protect firefighters from burnover. Prescribed fires are often used to reduce dangerous fuel levels and restore ecosystems; however, they can also be used for research and firefighter safety studies.

Figure 6.49

Figure 6.50

6.5.2 Wildland Fire Spread

Ian Grob
U.S. Forest Service

One of the greatest challenges in wildland fire is accurately predicting its spread. Experiments conducted at the US Forestry Service, Missoula Fire Sciences Laboratory, used huge arrays of small cardboard "tines" to reproduce wildland fire spread in a 3 m × 3 m diameter fire wind tunnel with a homogenous fuel source. The experiments showed that the flame structure is quite complex, with "peaks and troughs" formed in the flame front that allowed "bursts" of flame to push hot gases forward and preheat or ignite fine fuels far ahead of the mean flame front. An array of fine thermocouples, shown in one pane, was used to track the flame and these forward bursts. This knowledge has been used to understand the flame spread process and inform model development.

Figure 6.51

Reference

1. M. A. Finney, J. D. Cohen, J. M. Forthofer, S. S. McAllister, M. J. Gollner, D. J. Gorham, et al., Role of buoyant flame dynamics in wildfire spread, *Proceedings of the National Academy of Sciences* 112, 32 (2015), 9833–9838.

Figure 6.52

6.5.3 Firebrands

Luis Sinco

Los Angeles Times

Firebrands, the small burning embers that break off structures or vegetation during wildland fires, present an enormous hazard to life and property. They can create a "shower" that ignites structures and wildlands far from the original fire front.

In this photo a firefighter is seen walking through flying embers as flames from the Sayre Fire burns in the Granada Hills section of Los Angeles. The fire originated in Sylmar and jumped two freeways as it burned west into Granada Hills.

Figure 6.53

6.5.4 Firebrand Generation

Samuel Manzello
NIST

Firebrands, small burning embers that are responsible for ignition of many homes during wildland fires, are produced from a 5.2 m Douglas-Fir tree under no applied wind field.

Reference

Manzello, S. L., Maranghides, A., Shields, J. R., Mell, W. E., Hayashi, Y., & Nii, D. (2009). Mass and size distribution of firebrands generated from burning Korean pine (*Pinus koraiensis*) trees. *Fire and Materials*, 33(1), 21–31.

Figure 6.54

6.5.5 NIST Dragon

Samuel Manzello
NIST

Sayaka Suzuki
BFRL

A unique experimental apparatus, the NIST Dragon, has been developed to reproduce a "shower" of firebrands that a structure may experience during a large wildland fire. The firebrands shown here produced smoldering ignition in the shredded hardwood mulch bed (foreground). This transitioned to flaming ignition in the mulch bed, and the cedar fencing has just ignited in this image. Shredded hardwood mulch was intended to represent various fine fuels that are often found adjacent to fencing in the Wildland–Urban Interface (WUI). The dangers of wind-driven firebrand showers are apparent.

Reference

Manzello, S. L., Park, S. H., Suzuki, S. S., Shields, J. R., & Hayashi, Y. (2011). Determining Structure Vulnerabilities to Firebrand Showers in Wildland-Urban Interface (WUI) Fires. *Fire Safety Journal*, 46.

Figure 6.55

6.6 Smoldering Combustion

6.6.1 The Great Fire of London

Delia Murguõa Gutierrez, Felix Wiesner, and Rory Hadden

University of Edinburgh

The picture was taken in 2016 during the celebration of the anniversary of the Great Fire of London of 1666. The remains of the buildings can be clearly seen in the flames, adding a sense of scale. There is a large vortex clearly visible in the flames. It is a good impression of how the Great Fire of London might have looked like to observers back in the day.

Figure 6.56

6.6.2 Smoldering Combustion

Francesco Restuccia, Nieves Fernandez-Anez, and Guillermo Rein

Imperial College London

Smoldering is a lower-temperature, flameless form of combustion that evolves heat due to oxidation directly on the surface of the fuel. This picture shows the smoldering of softwood biochar pellets (10 mm × 2 mm × 2 mm), which were produced from biomass. The glowing parts are smoldering components of the biochar, with combustion and intense heat being emitted. The darker parts are char, where combustion has not yet taken place. Biomass is becoming widely used for energy production, so laboratory-scale studies like these are being conducted to understand their behavior in larger scales and prevent accidental fires from developing in biomass transport and storage.

Figure 6.57

6.6.3 Firing Black to White

Francesco Restuccia, Egle Rackauskaite, Yuqi Hu, and Guillermo Rein

Imperial College London

When a solid fuel is burned, it chars, with combustion reactions progressively turning it into ash. For complete combustion, all the original fuel turns into ash. With incomplete combustion, different amounts of char and ash ratios are produced. The picture shows the transition of charcoal fuel (black) to ash (white), with the sample containing only char on the left, followed by each sample containing an ever-increasing amount of ash until the right-most sample contains pure ash from complete combustion.

Figure 6.58

6.6.4 From Biomass to Ash

Francesco Restuccia and Guillermo Rein
Imperial College London

Biomass is an important fuel for power generation, and its use is increasing. However, it is associated with many fire safety issues. As a porous reactive media, it is prone to self-heating, caused by oxidation reactions at low temperatures. Surrogate biomass (rice husks) reheated in an oven at temperatures above 150°C produces torrefied biomass. Initially the rice husk is torrefied, with increasing oven temperatures causing charring leading the biomass to ignition above 190°C, leaving only ash (top).

Figure 6.59

Figure 6.60

Figure 6.61

Figure 6.62

Figure 6.63

Figure 6.64

6.6.5 Thermal Decomposition and Intumescent Behavior of Rigid Poly(vinyl chloride)

Joshua D. Swann, Yan Ding, and Stanislav I. Stoliarov

University of Maryland

The images show gasification of rigid poly(vinyl chloride) (PVC) exposed to a radiant heat flux of nominally 30.9 kW/m^2 in a nitrogen atmosphere. Isolated from gas-phase combustion and solid surface oxidation, the PVC demonstrates highly charring and intumescent behavior during pyrolysis. These experiments, performed in a well-controlled gaseous environment, provide repeatable mass loss rate, spatially resolved temperature, and quantitative sample shape data, which is required to construct a comprehensive pyrolysis model. This series of images represents various stages of the PVC thermal decomposition and intumescent process at 50, 150, 300, 400, 600, and 700 s.

Acknowledgements

This work was made possible by funding from the National Science Foundation

Keywords:

Thermal decomposition; gasification; intumescencethermal transport

Author Index

Subject Index